PROPAGATION OF THE JAPANESE ORNAMENTAL CARP (KOI)

BY DR. SHMUEL ROTHBARD

Dr. Shmuel Rothbard
YAFIT (R&D) Laboratory
Fish Breeding Center,
Gan Shmuel 38810, Israel
Fax+972-6-320404

The Japanese ornamental carp (koi), in Japan known as **nishikigoi**, are color variants of the common carp *Cyprinus carpio* Linnaeus. Koi were developed about 200 years ago as ornamental fish by families of Japanese carp farmers in the Niigata Prefecture, Honshu island. In the 20th century koi became increasingly popular in Europe and America.

About 17 years ago koi were successfully introduced into Israel. At present, koi are produced in Israel by five fish breeding centers and exported worldwide under the trademark **MAG-NOY**. Exportation of Israeli koi is increasing, and today Israel is considered one of the world's leading countries in koi production.

The increasing importance of the koi industry was the main reason that encouraged the Israeli koi breeders to ensure reliable production by securing sufficient numbers of high quality fry for stocking and by steadily improving the methods and techniques for their propagation. At present, these demands are being met by adopting methods for the genetic improvement of the existing fish stocks, but also by application of artificially induced propagation. Propagation carried out under controlled conditions notably increased the initial survival and contributed to the improvement of the final quality of koi.

What are Quarterlies?

Because keeping Koi as pets is growing at a rapid pace, information on their selection, care and breeding is vitally needed in the marketplace. Books, the usual way information of this sort is transmitted, can be too slow. Sometimes by the time a book is written and published, the material contained therein is a year or two old...and no new material has been added during that time. Only a book in a magazine form can bring breaking stories and current information. A magazine is streamlined in production, so we have adopted certain magazine publishing techniques in the creation of these Quarterlies. Magazines also can be produced much cheaper than books because they are supported by advertising. To combine these assets into a great publication, we issued this Quarterly in both magazine and book format at different prices.

Distributed in the UNITED STATES to the Pet Trade by T.F.H. Publications, Inc., One T.F.H. Plaza, Neptune City, NJ 07753; distributed in the UNITED STATES to the Bookstore and Library Trade by National Book Network, Inc. 4720 Boston Way, Lanham MD 20706; in CANADA to the Pet Trade by H & L Pet Supplies Inc., 27 Kingston Crescent, Kitchener, Ontario N2B 2T6; Rolf C. Hagen Inc., 3225 Sartelon St. Laurent-Montreal Quebec H4R 1E8; in CANADA to the Book Trade by Vanwell Publishing Ltd., 1 Northrup Crescent, St. Catharines, Ontario L2M 6P5 ; in ENGLAND by T.F.H. Publications, PO Box 15, Waterlooville PO7 6BQ; in AUSTRALIA AND THE SOUTH PACIFIC by T.F.H. (Australia), Pty. Ltd., Box 149, Brookvale 2100 N.S.W., Australia; in NEW ZEALAND by Brooklands Aquarium Ltd. 5 McGiven Drive, New Plymouth, RD1 New Zealand; in Japan by T.F.H. Publications, Japan—Jiro Tsuda, 10-12-3 Ohjidai, Sakura, Chiba 285, Japan; in SOUTH AFRICA by Lopis (Pty) Ltd., P.O. Box 39127, Booysens, 2016, Johannesburg, South Africa. Published by T.F.H. Publications, Inc.

MANUFACTURED IN THE
UNITED STATES OF AMERICA
BY T.F.H. PUBLICATIONS, INC.

Quarterly

yearBOOKS,INC.

Dr. Herbert R. Axelrod,
Founder & Chairman

Neal Pronek
Chief Editor

yearBOOKS are all photo composed, color separated and designed on Scitex equipment in Neptune, N.J. with the following staff:

DIGITAL PRE-PRESS
Michael L. Secord
Supervisor
Robert Onyrscuk
Jose Reyes

COMPUTER ART
Thomas J. Ceballos
Patti Escabi
Sandra Taylor Gale
Candida Moreira
Joanne Muzyka
P. Northrup
Francine Shulman

Advertising Sales
George Campbell
Chief
Amy Manning
Director

©yearBOOKS,Inc.
1 TFH Plaza
Neptune, N.J. 07753
Completely manufactured in Neptune, N.J. USA

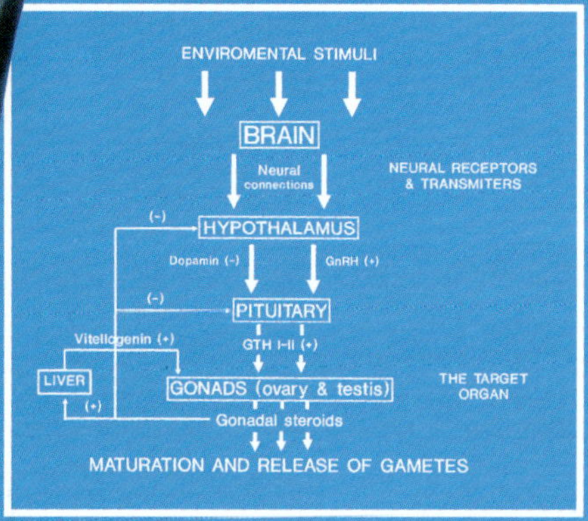
ENVIROMENTAL STIMULI
BRAIN
Neural connections
NEURAL RECEPTORS & TRANSMITERS
(-)
HYPOTHALAMUS
Dopamin (-)
GnRH (+)
(-)
PITUITARY
Vitellogenin (+)
GTH I-II (+)
LIVER
(+)
GONADS (ovary & testis)
THE TARGET ORGAN
Gonadal steroids
MATURATION AND RELEASE OF GAMETES

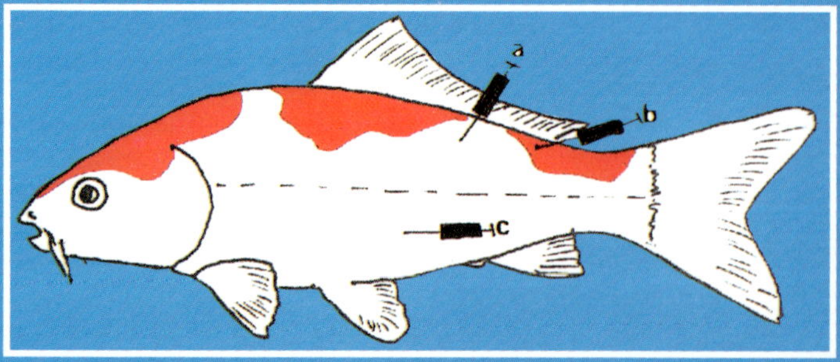
a
b
c

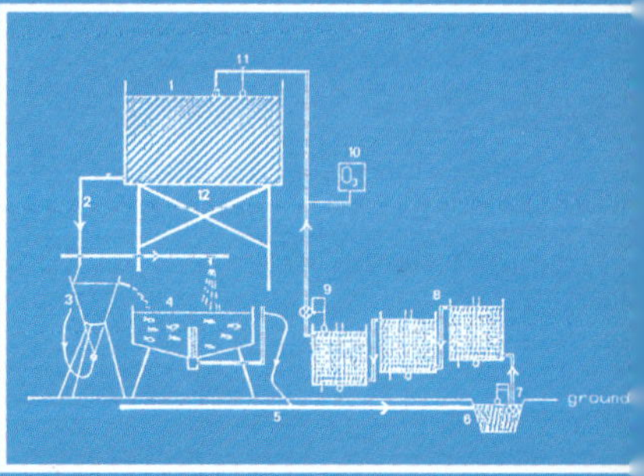

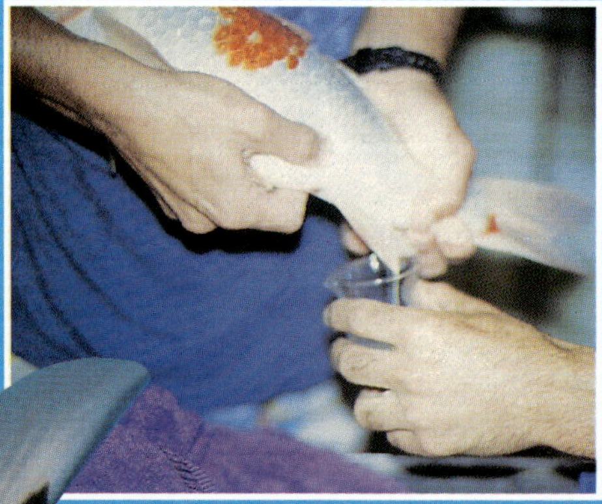

INTRODUCTION

The following criteria have been formulated and can be suggested for consideration to koi breeders in any koi propagation program:

1. Broodstock fecundity and related genetic inbreeding. Koi, like most other cyprinids, are extremely fecund fish and hundreds of thousands of eggs can be produced by each female. The result is that a small number of broodfish are capable of producing tremendous and, usually, sufficient numbers of fry. However, using only a limited number of spawners will result in increasing rates of inbreeding and genetic infirmity of the offspring. Cross-breeding of koi broodfish taken from different sources may avoid genetic inbreeding by the production of heterozygous and outbred progeny, thus sustaining a certain degree of hybrid vigor in traits that are desired by the koi hobbyist and koi grower.

2. Propagation methods. Koi can be spawned in earthen ponds using immersed spawning mats or pine-tree branches as substrates for the eggs. However, when planning field spawnings one should consider also the alternative, e.g. induced reproduction in a controlled environment. Recent advancements in the methods of propagation by means of spawning inducers (hormones known as GtH, GnRH, HCG, etc.), which are presently used in the form of injections or implants, improved the propagation of this fish and converted controlled reproduction into an art. When combined with the manipulation of environmental factors (temperature, photoperiod, etc.), these may efficiently control the maturation, release, and fertilization of the gametes.

3. Incubation of the eggs. In cases of pond spawnings, sticky koi eggs can be attached to water vegetation or to mats placed in cages (hapas), or directly to substrates immersed in the ponds. In indoor hatcheries, flow-through incubators of various shapes are used for the incubation of koi eggs that have been previously degummed. In each method, either in field or in indoor spawnings, water quality (temperature, oxygen, nitrogen compounds, etc.) must be monitored and controlled during all stages of egg development. Careful control and monitoring of the environment may have a remarkable effect on the survival of embryos and larvae.

4. Fry nursing. Small and properly constructed ponds that can be easily controlled, drained, and refilled with fresh water are preferable as koi nurseries. During nursing, the density of fry should be recorded by frequent weighing of representative fish samples to estimate the growth and survival of the fry. Too densely crowded nursing ponds will result in low survival rates. There is also another difficulty. At high stocking densities, the size frequency distribution of the fry tends to be asymmetric and the growth curve of the fish is skewed to the right. This characteristic is typical of fish populations and particularly of carps (one should not forget that koi are just a variety of carp). The "jumpers" were termed *Tobi-koi* by Japanese scientists, who were the first to investigate the phenomenon. That term was translated as *Shoot-carp* by Wohlfarth (1977). The *Tobi-koi* are fast growing fish that suppress the growth and development of smaller fish in the pond.

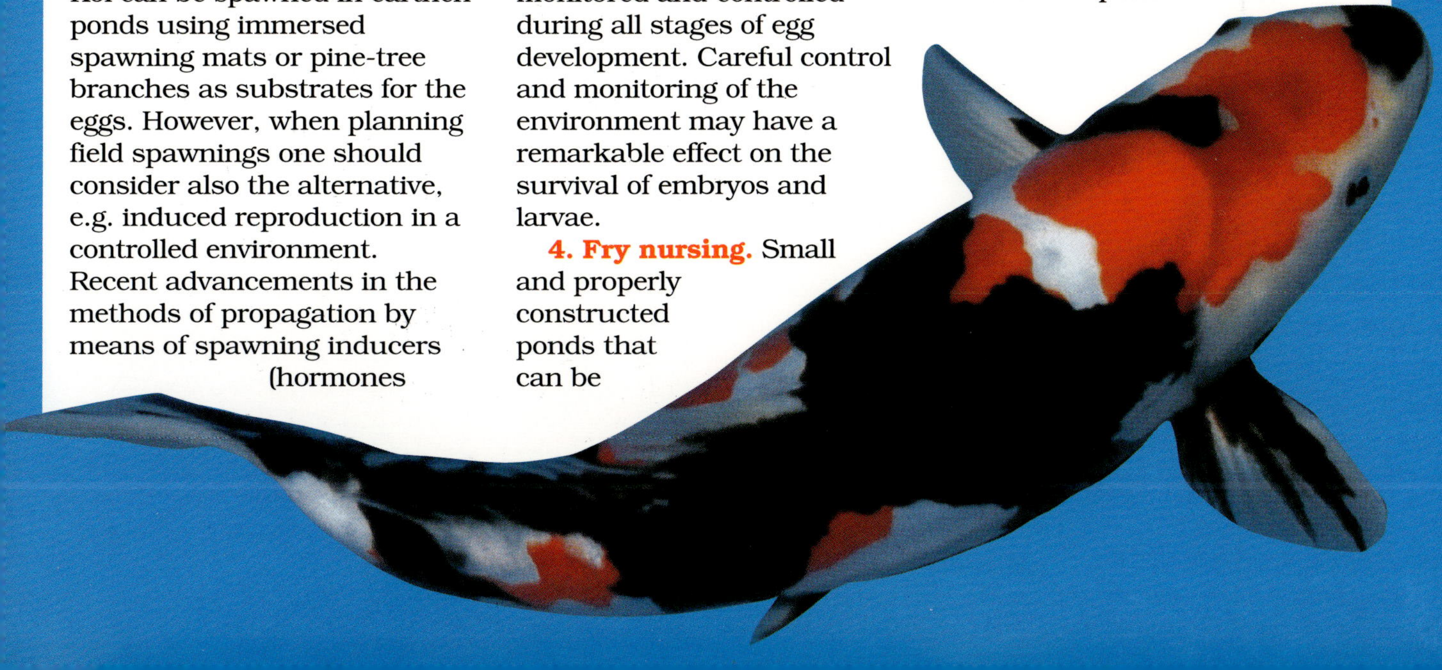

BROODSTOCK MAINTENANCE

HOW MANY BROODFISH ?

This, perhaps, will be the first serious question that every commercial koi breeder will face when starting a career with koi. The number of broodstock koi required for commercial reproduction undoubtedly depends on their spawning performance and their ability to produce salable fish that can be further selected for marketing. Therefore, it is clear that such considerations are related to, and have a direct impact on, koi broodstock quality. Nevertheless, methods of koi improvement are beyond the scope of this publication and have been already discussed in the March, 1995, issue of *Tropical Fish Hobbyist* magazine (Rothbard and Wohlfarth, 1995).

If we consider the quite primitive existing methods of culling and selection practiced by koi breeders, about 700,000 eggs are required to produce 50,000 koi of good quality. It means that only about 7% of the fish possess marketing potential and can be sold at a wide range of sizes. This fantastic number of eggs can be easily obtained from three medium size (ca. 2 kg) koi females. Consequently, a minimal broodstock comprised of 70-100 females and about the same number of males is needed in a commercial koi hatchery that has plans to produce about 3 million salable fish. In addition to that, and according to the Israeli (MAG-NOY) experience, about 25% of the koi broodstock has to be replaced each year due to unpredicted losses. When spawnings are carried out in earthen ponds, larger numbers of broodfish are required. In **Table 1** some of the differences between propagation methods applied indoors and in the spawning ponds are compared. Undoubtedly, the yield of fry produced indoors in the hatchery, being more efficient, is therefore much higher.

Figure 1. Koi broodstock.

MAINTENANCE OF KOI BROODSTOCK

Koi broodstock must be comprised of healthy and sexually mature fish **(Fig. 1)**. In koi, similarly as in common carp, temperature plays an important role in gonadal recrudescence and ovulation (Horvath, 1986). Koi females start the ovarian recrudescence at the end of the spawning season (end of summer). This process is completed in the spring of the following year, when the female has already accumulated 1000-1200 **degrees x days (DD)**. In the subtropics (e.g. Israel), the calculation of the **DD**s required to achieve ovulation starts with a gradual increase of the water temperature to the level of 12-15°C. At the end of the **DD** accumulation, females become gravid and are prepared to ovulate.

In a temperate climate, the process of koi maturation is extended to 3-4 years, the males maturing one year earlier than females. Koi broodfish sexually mature at the age of three years and, therefore, koi (and common carp) broodfish must accumulate at least 2000 **DD** to ovulate. Calculation of the **DD**s starts in the spring, at the third year of their life (Horvath, 1985, 1986) when most of the broodstock has already reached sexual maturity.

Showa Sanshoku.

As a routine procedure, after the spawning season has been accomplished, koi spawners are restocked into ponds, where they are maintained until the next spawning season.

In the subtropics, spawners that were induced to reproduce at the beginning of the season (February-March) can be induced to spawn once more, at the end of summer (July-August). These are the so-called late spawnings. In late spawnings the fish tend to produce fewer eggs than in the first spawning. The broodstock is prepared for the forthcoming season at the end of winter, before pond temperatures start to increase. This is the right time to segregate the fish according to their sexes and performance. The males and females are stocked into separate, small (0.2-0.3 ha) ponds. Separation of the sexes prevents uncontrolled reproduction. While stocking the spawners, attention should be paid so as not to exceed the broodfish biomass of 1500-2000 kg/ha. The broodstock pond must be equipped with an efficient aeration and a reliable source of good quality water to ensure good maintenance of the fish. Koi spawners at the age of 2-4 years (1.5-4.0 kg) are the most suitable fish for handling and propagation. **Table 2** presents some biological data concerning koi propagation.

TABLE 1

Comparison of hatchery and pond methods of koi propagation in a subtropical climate. The data considers production capacity of 500,000 salable koi of various sizes.

DATA COMPARED	HATCHERY	POND
Equipment	Special facility tanks, installations, etc.	Properly constructed ponds (earthen/concrete)
Broodstock	30 females (2kg) 20-30 males	200 females (2kg) 150-200 males
Ovulation	Totally Controlled	Environment-dependent
Spawning inducer	Necessary	Used but not required
Labor	2 persons	3-5 persons
Water system	Controlled/recirculated	Outdoor open pond
Egg treatment	Degumming required	None
Incubation	Conical incubators	Substrates for eggs
Larvae per kg of female	100,000	10,000
Primary nursing	7-10 days indoors then in small ponds	Continued in spawning pond to 1g
Survival to fry at 1g	50% (50,000)	35% (3000)
Know-how	Biological & technical skills	Pond management experience

Taisho Sanke

FEEDING KOI SPAWNERS

In general, it is agreed among hatchery experts that healthy fish come from healthy fry, which come from healthy larvae, which come from healthy eggs, which are produced by **healthy broodstock**. Koi broodfish require special feeds as well as a special feeding regime. Since many food formulas are advertised and available on the market, we will restrict this part of the discussion to general ideas and considerations.

Koi broodstock require a balanced diet with an adequate proportion of nutrients and vitamins. As in other fishes, koi spawners must not grow too much and not accumulate excessive fat. The spawners need to be healthy and in perfect performance to overcome stressful handling and other situations during the spawning cycle. They must have enhanced disease resistance to be able to recover after vulnerable treatments that involve stressful transportation, hormone injections, and stripping. It should be always kept in mind, however, that production of the next generation of eggs starts immediately after spawning, therefore spent broodfish must be maintained in perfect conditions until recovery.

At springtime, due to the relatively low biomass of the broodfish, the broodstock storage ponds are rich with zooplankton and natural nutrients. However, to enhance gonadal development and to avoid accumulation of fat, protein rather then carbohydrate-rich supplementary feeds are preferable for feeding the broodstock (Hepher & Pruginin, 1981). Normally, spawners can be fed year round with pellets containing 20-30% protein and a vitamin mix at the rate of ca. 1% of their biomass. This type of food contains sufficient proportions of carbohydrates/fat (energy) and protein. The last are very important for the recovery of koi spawners. It will also hinder too fast an increase in their body mass. The immune system of koi spawners will be enhanced by vitamins and other natural additives in the feed. The recovery period of broodstock is extremely important and has an enormous impact on the development of the gonads. Ensuring good recovery of broodstock is like a promise that the next year will be a success. One should adapt the feeding schedule according to the winter temperature fluctuations. At low temperatures (<14°C) koi stop active feeding. At springtime, 4-6 weeks prior to propagation, when the water warms-up, the regular food is replaced by protein rich (35-45%) pellets. The natural food, abundant in the spring, is an important factor which enhances gonadal development. In winter, at low temperatures, the feeding ratio can be reduced to 0.5% of the biomass.

TABLE 2

Reproduction data for koi broodstock.

Preferable age of broodfish	Koi 2-4 years old
Initial weight of spawners	>1 kg
No. of eggs/kg female	100,000
Spawning time (after 1200 DD)	At 20-26°C
Spawning methods	Hatchery or pond (see Table 1)
Spawning in nature	Shallow water + vegetation
Egg size (diameter); female age/size dependent	Pre-hydrated: 1.0-1.5 mm Activated: 1.5-2.5 mm
No. of pre-hydrated eggs/kg	600,000-1,000,000
Duration of incubation	2-4 $^{1}/_{2}$ days (temperature dependent)
Yolk sac absorption	2-3 days post-hatching
Free swimming fry (size)	After a week at 9-11 mm
Initial prey size (e.g. *Artemia*)	100-250 µm
No. of *Artemia* preyed/day/fry	150-300 ind/4-8 day old fry
Size of 3-4 week old koi fry	ca. 1g at a stocking density of 1,000,000/ha

Below: Showa Sanshoku.

Above: Taisho Sanke.

SELECTION & HANDLING OF KOI SPAWNERS

Due to their sensitivity at ripeness, koi broodfish are easily stressed by poor pond conditions; they are particularly affected by low oxygen levels and high biomass of fish. Careless handling at netting and transportation also have a negative effect on koi broodstock by suppressing the reproductive ability of the fish. All these requirements should be carefully considered in every activity with broodfish prior to inducement of spawning. Special attention should be given to avoid the excessive crowding of spawners in the net while netting the fish at the most sensitive stage of their gonadal development.

As soon as koi females have been netted they are examined for ripeness in the pond, without taking them out of the water. Each female is examined by a gentle palpation of her soft belly and papilla. Ripe females can be easily recognized by their enlarged and swollen genital (papilla) openings. The males are gently turned upside-down in the water, their ripeness being indicated by some drops of semen which are squeezed out.

For short distance (pond to hatchery) transportation, spawners selected from ponds can be tied in soft cloth sacks and placed in a tank filled with well oxygenated water. Fish transported in opaque sacks remain calm during handling. In the hatchery, koi females are weighed individually and, if possible, accommodated in separate holding tanks. The weight of the males can be just estimated and 3-5 koi males, 2-3 kg each, can be placed together in one 500-liter holding tank.

Figure 2. Selection of spawners in the pond.

Below: Taisho Sanke.

Above: Showa Sanshoku.

USE NARCOTICS TO AVOID STRESS IN THE SPAWNERS

To prevent injury and stress, the very sensitive koi spawners are tranquilized at each handling to make them calm at the time they are removed from the water. Individual weighings, administration of hormones, or stripping of eggs are carried out when the fish are anesthetized to the stage of a partial loss of equilibrium (Stages III-IV; **Table 3**). There are numerous anesthetics used in fish culture. We found the **Benzocaine™** derivative, ethyl-4-aminobenzoate, to be the most suitable for regular use. The chemical (available as powder) is dissolved in a small volume of ethanol (1 g/10 ml) and diluted at the concentration of 1g/10L water. Fish can be totally tranquilized in 2-3 minutes when placed in a bath containing Benzocaine™ solution.

At tranquilization, the fish must be watched and inspected so that immediately after the fish becomes calm and loses its equilibrium, it can be removed from the container for the planned

Figure 4. A hanging balance for the individual weighing of fish.

Figure 3. Another view of workers selecting spawners from a pond.

Below: Hi Utsuri.

Above: Hi Utsuri.

treatment. Soon afterward the spawner is returned, for recovery, to a well-aerated tank filled with fresh water, which may also include antibiotics. Thereafter the fish is returned to a storage pond.

OVARIAN BIOPSY OF KOI FEMALES

Induced reproduction in koi is commonly attempted at early springtime (February-April) in order to stock the ponds with young fry as early as possible. Nevertheless, at the beginning of the spawning season an examination of the broodstock by observation of external features alone is insufficient. At this time only a small part of the females in the broodstock will be ripe and able to respond to hormonal treatment. This will lead to frequent trials of unsuccessful spawnings. In order to avoid the waste of broodstock (e.g. improperly selected, hormonally treated, but non-ovulated females), a simple technique of biopsy is applied which facilitates selection of ripe females that are about ready for propagation.

A Tygon™ tubing of 4 mm (internal diameter) is inserted, via the genital open-

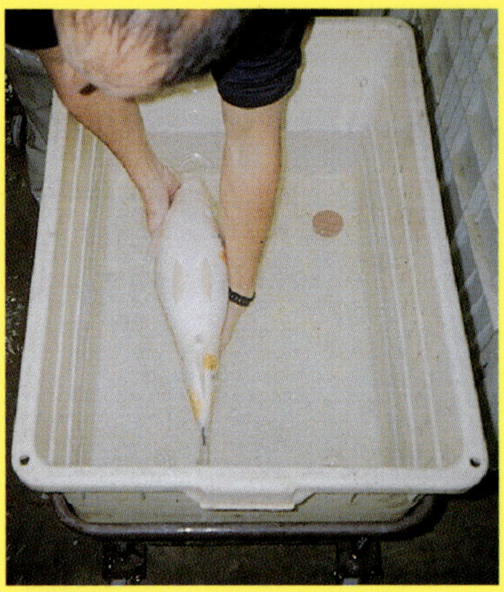

Figure 5: A female koi tranquilized and ready for stripping.

TABLE 3
Behavioral changes of fishes at anesthetization (modified after McFarland, 1960).

Stage of anesthesia	Terminology	Fish behavior
I	Normal	Response to external stimuli. Muscle tone and equilibrium normal.
II	Light sedation	Slight loss of response to external stimuli (visual and tactile).
III	Deep sedation	Total loss of reactivity to external stimuli (except strong pressure). Slight decrease of opercular rate.
IV	Partial loss of equilibrium	Partial loss of muscle tone. Response to very strong external stimuli. Swimming capability disrupted. Increase in opercular movements (rare).
V	Total loss of equilibrium	Total loss of muscle tone. Response to very strong pressure. Opercular rate slowed below normal.
VI	Loss of reflex reactivity	Total loss of reactivity. Respiratory and heart rates very slowed.
VII	Medullar collapse	Respiration ceases followed by death several minuts later.

The depth of anesthesia is affected by temperature, size of fish, and species.

Below: Hi Utsuri.

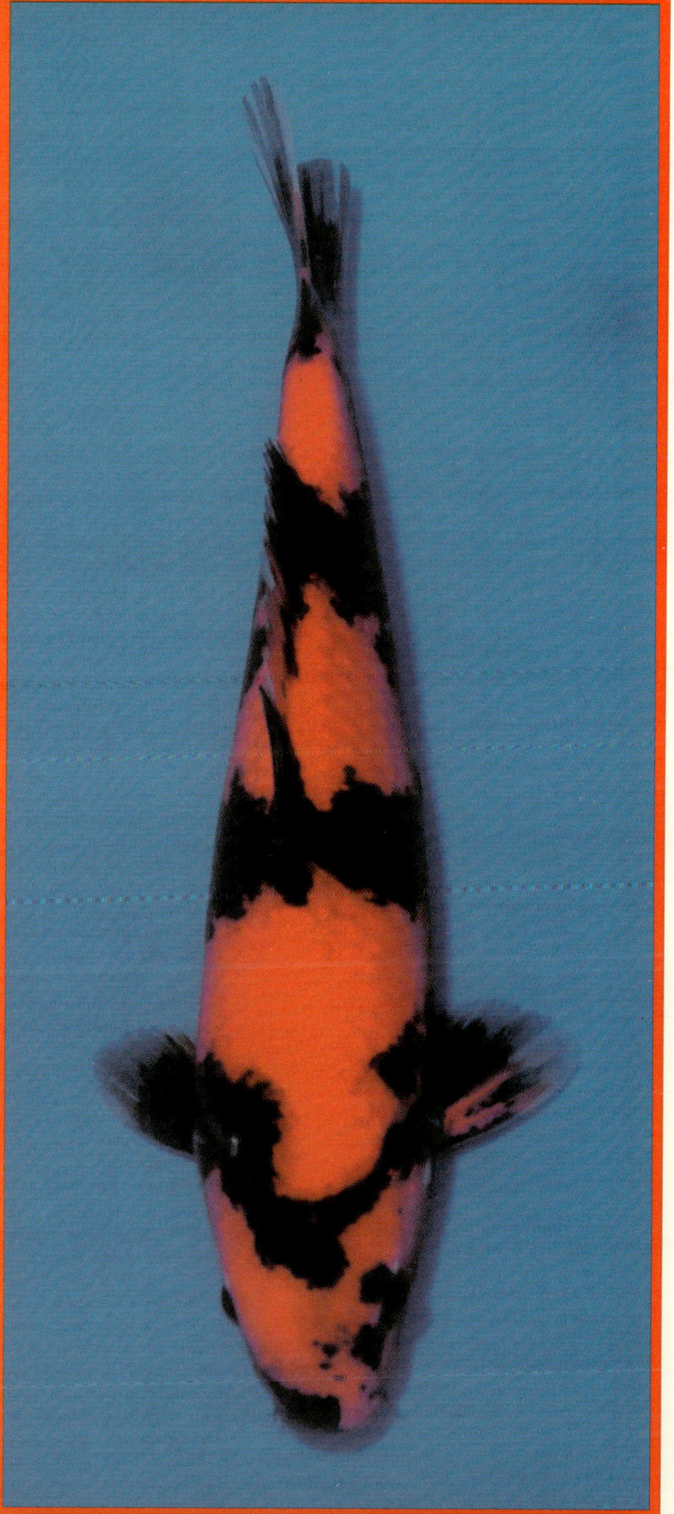

Above: Hi Utsuri.

ing and the genital duct, about 6-10 cm into the ovary and an ovarian sample of 40-50 eggs (follicles) is sucked out for examination under a binocular microscope. Since the koi (and carp) follicles are opaque, the oocytes in the sample are cleared in Serra's fluid (ethanol : formalin (40%) : acetic acid; 6:3:1 v/v/v). The opaque oocytes become translucent within 3 minutes and remain so for an additional 6-7 min., then they become opaque again. During this short interval the position of the germinal vesicle (GV) is determined in the sample. Successful spawning induction can be attained only in female carp in which at least 65% of the oocytes (GV) have already started their migration (GVM) from their central position toward the oocyte periphery (Yaron and Levavi-Zermonsky, 1986). Females with less than 60% oocytes that have not yet reached the stage of GVM, are returned to the pond and can be used on a later occasion.

Figure 7. Schematic presentation of oocyte development (see text for explanation).

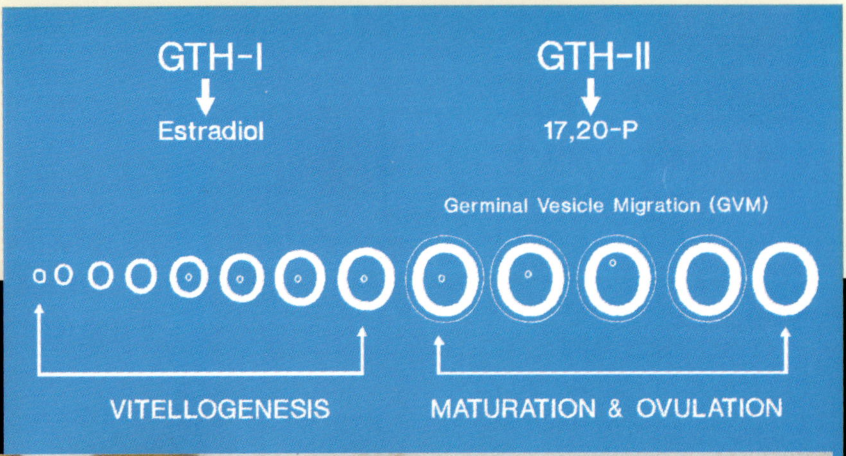

Figure 6. Below: Koi eggs treated with Serra for transparency. Note the position of the germinal vesicles in the egggs.

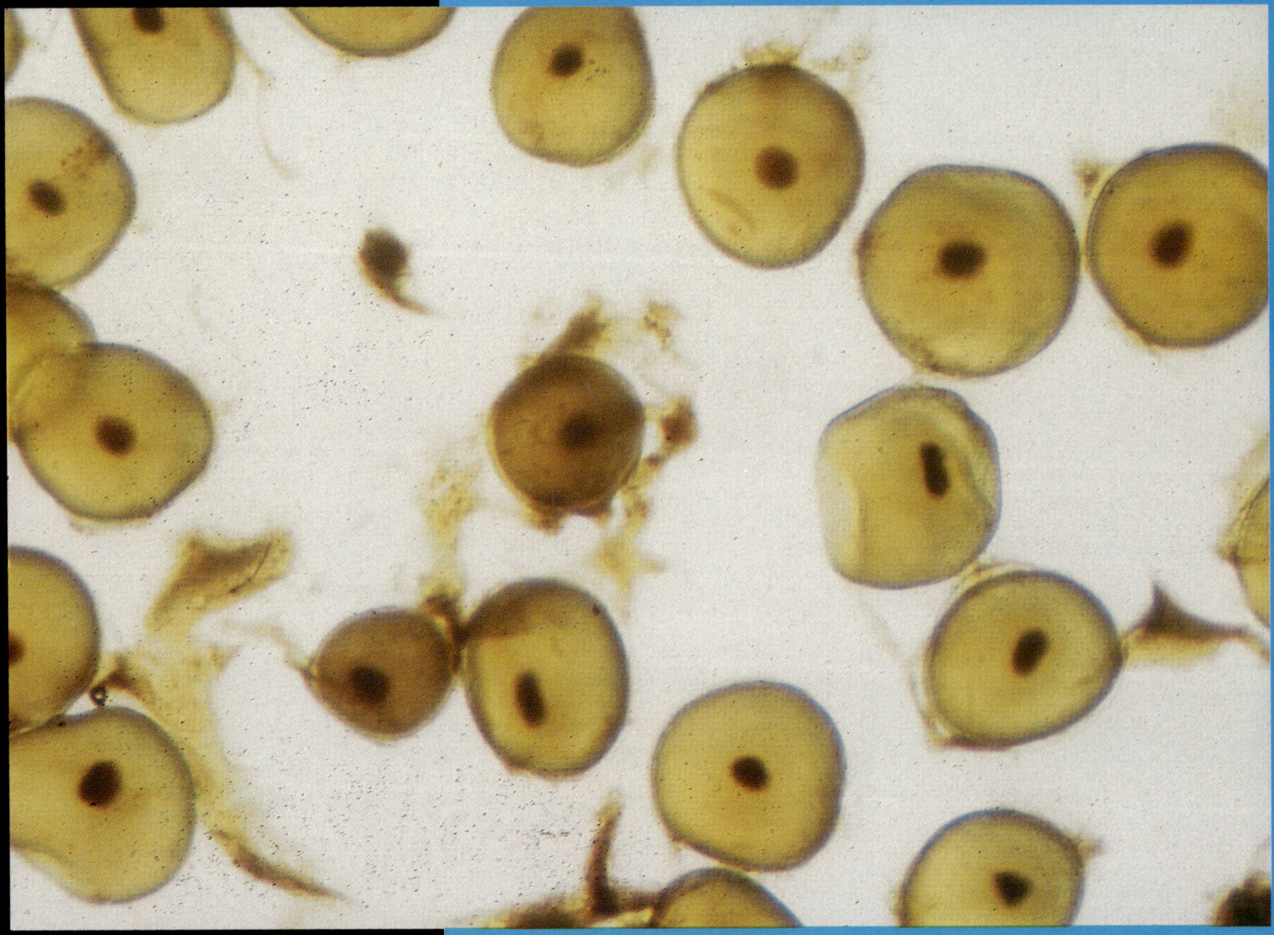

Below: Kohaku.

Above: Showa Sanshoku.

Below: Tancho.

Above: Koromo.

THE ENVIRONMENT/FISH INTERACTION

FISH REPRODUCTION AND THE ENVIRONMENT

The aquatic environment affects living organisms, including fishes, via a complex sensory (olfactory, visual, etc.) system. The system is stimulated either by environmental factors such as temperature, daylight, fluctuations in water level (river floods), social structure of the fish population, and the presence of aquatic vegetation, or by the so-called **biotic** and **abiotic** factors. These latter can be experimentally examined. They are divided into **Ultimate** and **Proximate Factors**, respectively. An important factor that may affect physiological processes associated with reproductive behavior and performance, is the presence of fish of another sex. Normally the fish release chemical substances (pheromones) into the aquatic environment to suppress or to enhance breeding behavior, or to suppress the growth of another fish. At spawning time sex pheromones will sexually attract fish of another sex. In an overcrowded fish pond, growth inhibiting pheromones will result in a stunted population.

The environmental signals transmitted via the aquatic medium are absorbed and translated by brain/neural receptors. The information is then transmitted through the blood and/or nervous system in the form of messengers (hormones) addressed toward internal target organs, including the sexual organs—the gonads.

For instance, environmental stimuli are assimilated by brain neuroreceptors as signals which promote the production of various hypothalamic hormones (hormones produced in the **hypothalamus**). Among these, the neural signals urge secretion of both the **Gonadotropin-Releasing Hormone (GnRH)** and its antagonist, **Dopamine (DA)**. Consequently, GnRH secretion stimulates production and release of gonadotropic hormones (**GtH-I** and probably the recently detected **GtH-II**) from the pituitary. These are targeted toward the gonads and trigger gonadal development, stimulate release of steroid hormones (E_2 and 17,20-P), and play an important role in vitellogenesis and oocyte maturation **(see p. 16)**. On the other hand, the GnRH antagonist, dopamine, suppresses the production and secretion of GtH, thus balancing and controlling the GnRH secretion. The steroids, namely **estrogens** in females, stimulate production of the yolk lipoprotein (**vitellogenin**) in the liver, while the **androgens** in males stimulate production of sperm in the testes. The estrogen (Estradiol-17) acts also as an antagonist toward the hormones, which are released from the hypothalamus and the pituitary via a complicated regulatory system of negative feedback. Schematic presentation of the environment-brain-hypothalamus-hypophysis-gonads physiological axis is shown in the chart on p. 20.

We can summarize this part with a general conclusion: The reproduction of fishes (and other vertebrates) is a combined act of coordination between the nervous and endocrine systems. The neural system prevails at the beginning of the reproductive cycle, whereas the endocrine (hormonal) activity predominates in the final stages of sexual maturation and the production of gametes, ovulation, and spermatogenesis.

THE GONADS

As in all fishes, sexual maturation of koi spawners is temperature dependent (Horvath, 1986). In fact, the maturation is affected by the time during which the broodfish are exposed to temperatures higher than 16°C. Such exposure occurs at springtime and permits accumulation of the required amount of **DDs**. In the subtropics (e.g. Israel) the koi sexually mature at the age of one year, while in temperate climates (e.g. Japan and Europe) sexual maturation is postponed and koi spawners mature at the age of 2-3 years. There is some experimental evidence that carp (koi) maturation can also be mediated by fluctuations of the day length (namely the photoperiod). It was shown that the photoperiod affects diurnal changes in the levels of GtH-II during ovulation and may affect the effectiveness of hormonal treatments.

There is a clear association between ovarian maturation and seasonal or daily fluctuations in hormone levels. In a study carried out on common carp that were maintained under Israeli conditions (Yaron & Levavi-Zermonsky, 1986), this association was investigated in detail. Apparently, maturation of ornamental carp (koi) is identical to that of common carp. In the carp study, samples were collected monthly during the entire year. Hormone levels were determined in blood samples as well as in pituitary glands, using sensitive radioimmunoassays (**RIA**). At the same time samples of female gonads were examined in order to calculate the **Gonado-Somatic-Index** (**GSI**), e.g. the ratio between the gonad and body weight (gonads weight x BW^{-1} x 100). Also, follicle size was examined in samples collected from excised ovaries of sacrificed fish.

The following conclusions have been drawn from the study and they may have a direct and practical implication to koi breeding, with special respect to propagation carried out in the northern hemisphere:

• During the spawning season, two distinct populations of oocytes can be detected in samples of oocytes examined microscopically. At the beginning of the spawning season (February-March), large pre-ovulatory oocytes (1.1 mm) and small size follicles (>0.15 mm) are found in the ovary. The small follicles (stages I-III of oocyte development; the ovogonium) will mature in the next year and the large size preovulatory oocytes are about to ovulate. However, in late summer, at the end of the spawning season, only small follicles remain in the ovary.

• The size of oocytes is closely associated with the gonado-somatic index (GSI). Large oocytes (stages IV-VII of oocyte development; the vitellogenic phases) are found in the ovaries together with an elevation of GSI to its highest values (ca. 10.0 in May). It is typical to and normally occurs in the middle of the spawning season. When the ova reach the last stage of development (VII) they may remain for several months as dormant eggs, ready to ovulate in favorable conditions.

• Two gonadal hormones, Estradiol-17ß (**E_2**) and testosterone (**T**), and the gonadotropic hormone (**GtH**) released from the pituitary into the blood system, were detected and measured. They showed a significant peak at the onset of spawning and an additional rise at the end of summer (September). Additional peaks of these hormones, which appear in late summer, probably indicate commencement of a consecutive process of vitellogenesis (accumulation of yolk lipoproteins in the ova). The reason for the repetitious accumulation of vitellogenin is to generate ova that will become mature for the next year's propagation.

Figure 8. Schematic presentation of the exogenous and endogenous factors affecting and controlling reproduction in fish: GtH, gonatotropic hormone; GnRH, gonadotropin releasing hormone; (+), a positive and (-), a negative feedback.

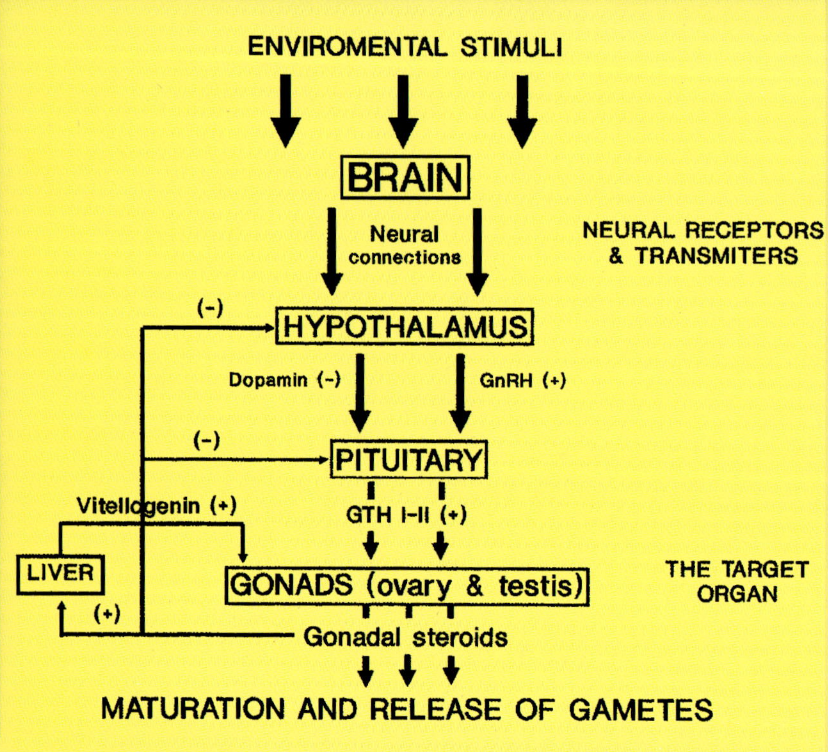

Below: Shiro Utsuri.

Above: Taisho Sanke.

INSTALLATION OF A KOI HATCHERY

THE WATER SYSTEM

A recirculating water system, which makes possible the control of temperature, water quality, hydrostatic pressure of the dissolved gases, and oxygen saturation, is a must in a koi hatchery. It is a prerequisite for efficient incubation of eggs as well as for the primary nursing of larvae and small fry. Normally, high hydrostatic pressure in water pipes (4-8 atms) is the main reason for supersaturation with atmospheric gases that tend to generate air bubbles. These adhere to the incubating eggs, which then become buoyant and rise to the water surface, to finally be drained from the incubator via overflow. Water supersaturated with high pressure gases is also fatal for the eggs and larvae. The small larvae and fry suffer from gas bubble disease, due to the diffusion of excessive amounts of gases into their body from the water (see the section on gas bubble disease). A recirculating water system based on a continuous supply of good quality water and constant but low water pressure, may prevent hazardous and unexpected catastrophes.

The water system of the Gan Shmuel fish hatchery is presented here as a model. The recirculating water system has a capacity for the incubation of about 10 million eggs in one cycle of reproduction (see figure p. 24). It includes a 3-4 m^3 upper level storage tank for the filtered water situated at 5-8 m above the hatchery ground level. The tank is supported by an outer source of well water, controlled by a floating valve at the inlet. The most important role of the storage tank is to secure a continuous and uniform supply of good quality water at low hydrostatic pressure. There is an option to install a thermostatically controlled heat-exchanger inside the upper tank in order to control the temperature-dependent latency of spawners during hormonal treatment. The heater also makes it easier to control the duration of incubation and the hatching of the eggs.

From the upper storage tank the water flows by gravity into the hatchery, into the conical incubators and larvae tanks. Then the water is drained into the lower level water accumulation tank, which serves also as a settling tank for solid wastes. From this tank water is pumped into a row of biological filtration tanks, the first being on the highest level and the last at the lowest level. The filters consist of 1 m^3 tanks filled with small size (3-5 mm) dolomite or tuff gravel. These have a double function, they serve for mechanical filtration but soon develop as biological filters. The water flows via siphons from one to the next, lower level filtration tank. The quality of filtration depends directly on the number of filtration units, the more filtration tanks the better quality of water. In the Gan Shmuel hatchery, the filtration system consists of 10 units, comprising together a 10m^3 biological filter.

The filtered water is raised to the upper storage tank by a buoy-controlled electric pump. There is a possibility of placing a disinfecting unit, either a battery of UV-tubes or an ozonizer (which injects strongly oxidizing molecules of O_3 into the water), between the last filtration unit and the upper storage tank.

HOLDING KOI BROODFISH INDOORS

Most fish hatcheries are equipped with bio-filtration, with the more advanced water systems also being thermoregulated. Such a system prevents the waste of water and energy for heating, but also provides water of low pressure in the pipes. Thermoregulation is important for controlling and therefore to predicting the timing of ovulation.

Koi spawners introduced into the hatchery can be placed in 0.5-1.2m^3 holding tanks supplied with running, good quality, well water. Holding tanks are disconnected from the main hatchery water system until the fish introduced from outside the hatchery have been disinfected. Large broodfish are placed in tanks of appropriate size. Each tank is provided with a continuous water supply and aeration, and is shaded and protected by netting to keep the fish calm and to prevent their jumping out and possible injury.

Below: Kohaku.

Above: Showa Sanshoku.

Fish are acclimated to the hatchery conditions for one to two days before being treated against ectoparasites that can be introduced from the pond. The fish are disinfected for 5-6 hours with a combined solution of 0.2 ppm malachite green, as recommended for application in fish hatcheries by Alderman (1985), and 30 ppm of technical grade (37%) formalin. After being disinfected, the tanks are reconnected to the hatchery water system. In some hatcheries fish are treated with various concentrations of sodium chloride (NaCl; cooking salt) solutions, which rids the skin of a wide range of bacterial and protozoan ectoparasites. For large koi spawners a 2.5% (25,000 ppm; 25g NaCl/L water) solution bath, applied for 10-15 minutes, is effective against *Monogenea*, various parasitic copepod larvae (anchor worm, *Lernaea*), and various fungi (Kabata, 1985).

Spawners are not fed in the hatchery. This avoids the presence of feces in the tanks and during the stripping of eggs and sperm. Hormone treatment can be initiated the next day or postponed for several days. When separated from males, ripe koi females may be kept in the hatchery, at low water temperatures (18-22°C), for several weeks prior to the initiation of spawning. Koi males with running milt may be maintained in similar conditions all year round and stripped several times, at 2-3 week intervals.

Figure 9. Schematic plan of a small hatchery for koi egg incubation and primary nursing. Arrows show the direction of water flow

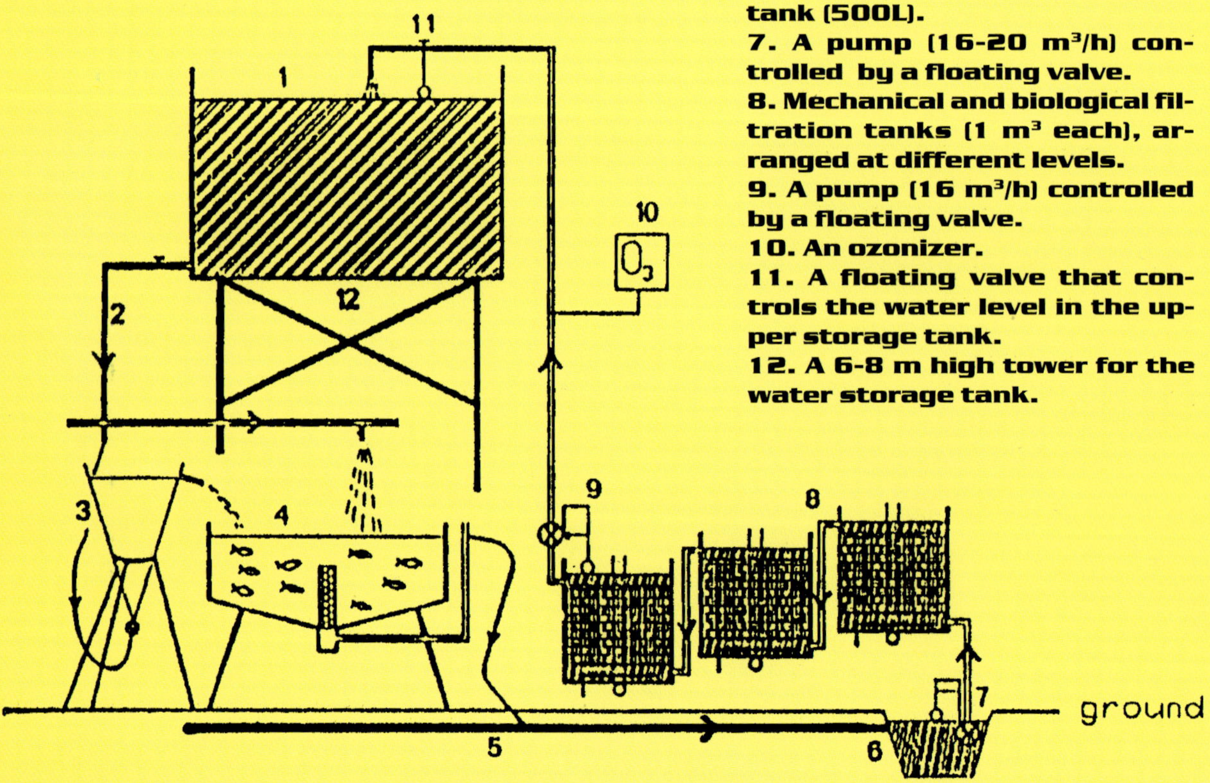

1. Water level in the upper water storage tank (3-4 m^3).
2. Plastic pipe (2") that supplies water to the hatchery.
3. Conical incubator (50-80L) for koi eggs.
4. Circular tank (1.5-2.0 m^3) with sloped bottom and a bottom drain (4") for koi larvae.
5. A drainage pipe (6") that carries water toward the lower water accumulation tank.
6. Lower water accumulation tank (500L).
7. A pump (16-20 m^3/h) controlled by a floating valve.
8. Mechanical and biological filtration tanks (1 m^3 each), arranged at different levels.
9. A pump (16 m^3/h) controlled by a floating valve.
10. An ozonizer.
11. A floating valve that controls the water level in the upper storage tank.
12. A 6-8 m high tower for the water storage tank.

Below: Kujaku.

Above: Sumi-Nagashi, a Kawarimono.

Below: Asagi.

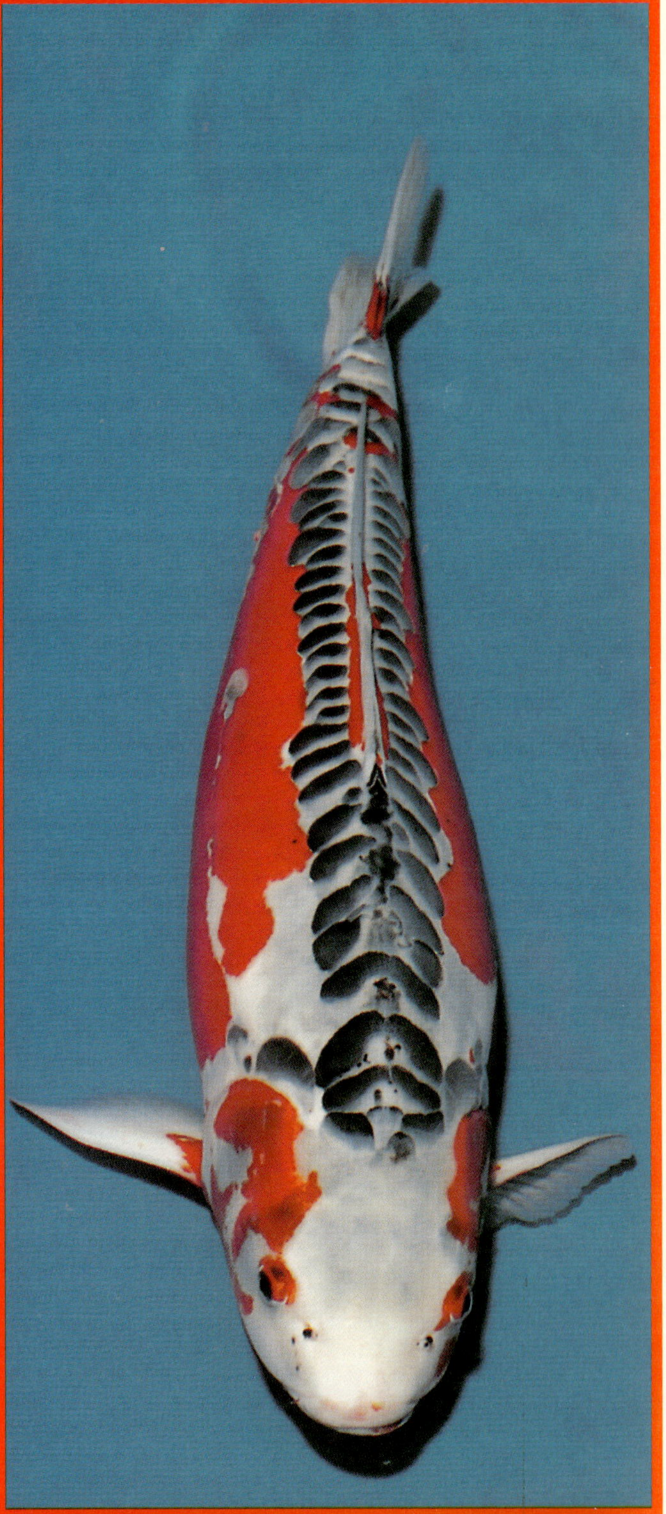

Above: Shusui.

INDUCED REPRODUCTION OF KOI

HISTORY

In the middle of the 40s a short and interesting article was published by a Brazilian scientist, Dr. von Ihering (1937), who wrote:

"....I may call the attention of the American readers to a method which, as far as I know, has been practically applied only in Brazilian fish culture. I refer to the forcing of spawning in fishes in captivity when they are unwilling to reproduce. We.....are able to obtain the needed eggs from these fishes only by pituitary injections."

The Brazilian scientist started the work in 1934 with South American fishes (Characidae). He successfully induced spawning in fishes that were injected with homogenized (in saline) and extracted pituitaries collected from heterologous and mature donor fishes. At the same time, a Russian fish culturist, N. L. Gerbilsky, succeeded in inducing spawning in fish of the sturgeon family (*Acipenser stellatus*), using extracts from freshly collected pituitaries of homologous species. Today, it can be undoubtedly said that von Ihering's short communication and Gerbilsky's work started a new era in fish culture and particularly in fish propagation. Controlled reproduction of fishes, with the aid of externally administered hormones, became a worldwide and routine practice in numerous fish hatcheries and in most cultivated fishes.

HYPOPHYSATION

Hypophysation is a technique aimed at inducing final maturation and ovulation by increasing the levels of GtH in a ready-to-spawn fish. This is attained through the elevation of endogenous GtH levels by the administration of pituitary extract that contains GtH from an external source. Until recently, the common technique of hypophysation involved the use of crude pituitary glands collected during the spawning time from freshly decapitated fish. The collected pituitaries were preserved either in ethanol or dried in acetone until use. It was believed that pituitaries collected from females during the spawning season are more potent in comparison to the glands collected from males. The glands were blotted and homogenized in a saline solution (0.9% NaCl) prior to injection. Usually, pituitaries were used on a "weight to weight" basis, and pituitary material collected from a 1 kg fish served to induce spawning in broodfish of a similar body mass. Practically, the heavier broodfish required an adjusted quantity of pituitaries. The total dosage required was divided into two portions. For a 1 kg fish, the first dose of 0.2 gland diluted in saline solution was given for priming and 12-14 hours later the resolving dose of 1.0 gland was injected. Since the ovulation is temperature dependent, spawning occurred 8-14 hours later. Males were injected only once, with 50% of the female's dose and 12-20 hours prior to stripping (**Table 4**).

In general, spawnings induced with crude pituitary extract were satisfactory and, if carried out by a skilled technician, usually resulted

TABLE 4

The working schedule of hormonal treatment to induce spawning in koi, as carried out at a water temperature of 24°C.

THE TREATMENT	SPAWNERS (DOSE) FEMALE	MALE	TIME BEFORE STRIPPING
CPE or crude pituitary extract			
1st injection	20%	-	ca. 18 hours
2nd injection	80%	50-60%	ca. 9 hours
GnRH (and analogues)			
Only 1 injection	100%	50-60%	ca. 14 hours

in 50-70% ovulated fish. However, in case of a non-success it was almost impossible to ascertain whether the failure was caused by a wrong selection of a spawner or impotent (low GtH content) pituitaries. A fundamental study on induction of spawning and fluctuations of hormones (GtH and steroids) during induction of spawning of common carp, was carried out and reported in detail by Yaron & Levavi-Zermonsky (1986) and Levavi-Zermonsky & Yaron (1986). As a result of the work, a GtH-calibrated carp pituitary extract (CPE) was developed. The CPE was commercially manufactured and used in Israel until recently when it was successfully substituted for by a GnRH analog. A similar technique of hypophysation using CPE as a spawning agent was also successfully applied in the color variant of common carp, the *nishikigoi* (koi), as well as in many other fishes.

Satisfactory spawnings can be attained when crude pituitaries are used after storage in pure ethanol for more than one year. The CPE has a longer shelf-life and can be stored in a home refrigerator at 4-5°C (**avoid deep freeze !!!**) for years. A high rate of ovulation and spawning response was achieved in koi injected with CPE that was stored for 3-4 years.

THE HYPOTHALAMIC APPROACH TO INDUCED REPRODUCTION

The relatively novel hypothalamic approach (developed at the early 80's) to induce spawning is based on external intervention in the hypothalamus-hypophysis axis. This is carried out by the administration of exogenous mammalian or bony fish **gonadotropin releasing hormone (GnRH),** a peptide of 10 amino acid sequences (decapeptide) or its synthetic analogs. When combined with dopamine (DA) antagonists, it hinders the DA effect and does not prevent the surge of the level of endogenous GtH. There are some synthetic GnRH's available on the market. One of these is the super-active GnRH analog (sGnRH-a) which includes the DA antagonist metochlopramide manufactured in Israel and distributed under the trade name **DAGIN**.

There are some advantages in the application and use of sGnRH-a:

1. It stimulates an increase of the endogenous GtH levels and their release into the circulatory system.

2. Since GnRH is a relatively short (10 amino acids) peptide, it does not provoke any immunological side effects in the treated fish.

3. The dosages are relatively small (10-50 µg/kg spawner) and are given as one portion. This reduces the stress of the fish to a minimum (**Table 3**). In some (marine) fishes the hormone is applied in the form of silastic implants, with a sustained and slow release of their contents into the blood stream.

4. The differences between the peptides are very small and therefore GnRH (and the sGnRH-a as well) from one species can be successfully used as a spawning inducer for many fishes.

In the Gan Shmuel (MAG-NOY) Fish Breeding Center (Israel), DAGIN is extensively used to spawn koi and other fishes (Drori, *et al.* (1994). As with other synthetic GnRH, DAGIN can be stored for long periods of time (at least five years) in the originally vacuum-packed vials.

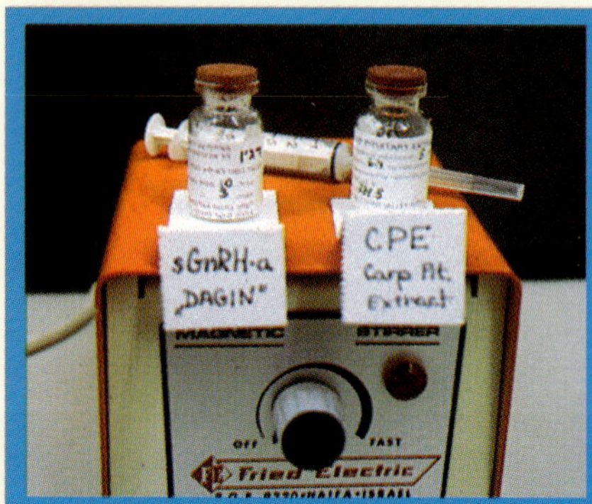

Figure 10. Vials containing the CPE and the sGnRH-a hormones.

THE HORMONAL INDUCTION OF SPAWNING

To induce ovulation, the total portion of either crude pituitary homogenate or CPE are injected in two doses at different times. The first (priming) injection stimulates migration of the GV toward the animal pole prior to maturation and the ovulation of oocytes. The role of the priming dose is to induce the gonads to pass into the pre-ovulatory stage. The priming dose comprises 20% of the total portion and is injected 12-14 hours before the second injection. The second (resolving) dose, which contains about 80% of the total portion, is injected 10-12 hours prior to the planned stripping. Since the fish response to the entire treatment is temperature-dependent, its timing can be controlled (extended or shortened) by manipulation of the

Below: 'B' Ginrin.

Above: Bekko.

Below: Hikarimono.

Above: Doitsu.

water temperature. At 24°C ripe koi females will ovulate 8-10 hours after the second injection of CPE (**Table 4**).

Ovulation induced with the sGnRH-a (+DA antagonist) stimulates release of endogenous GtH from the pituitary. The DA-antagonist hinders DA receptors and, therefore, results in enhanced levels of endogenous GtH. Similar to CPE, sGnRH-a activity is also temperature dependent. Koi, as other fish, ovulate earlier after hormonal treatment if maintained at higher temperatures (22-26°C). The latent time (latency) between the resolving dose of CPE, or the sole sGnRH-a injection, is presented in (the graph opposite). The final CPE dose or the sole dose of sGnRH-a results in final oocyte maturation and ovulation in the koi female, but also increases milt production in koi males. Since the processes are temperature-dependent, the latency should be carefully considered to avoid too late stripping and thus over-ripening of the eggs.

Since the koi spawners are tranquilized before the treatment, the hormones can be carefully injected into the dorsal musculature behind the dorsal fin or intraperitoneally (see figure opposite). Usually a volume of 0.5-1.0 ml of saline dissolved hormone is injected per kg of spawner. To inject the fish, one of the scales is carefully lifted and the syringe needle inserted into the muscle. After the injection, and before removal of the needle, the muscle is gently massaged with a finger to discharge the fluid inside the body.

Figure 11. Curves showing the temperature (°C) dependent latency (h) between either the resolving dose of CPE or the dose of sGnRH-a and the ovulation in koi females. Each point represents the average latency of females. Numbers attached to each point indicate recorded number of females that ovulated at given temperature.

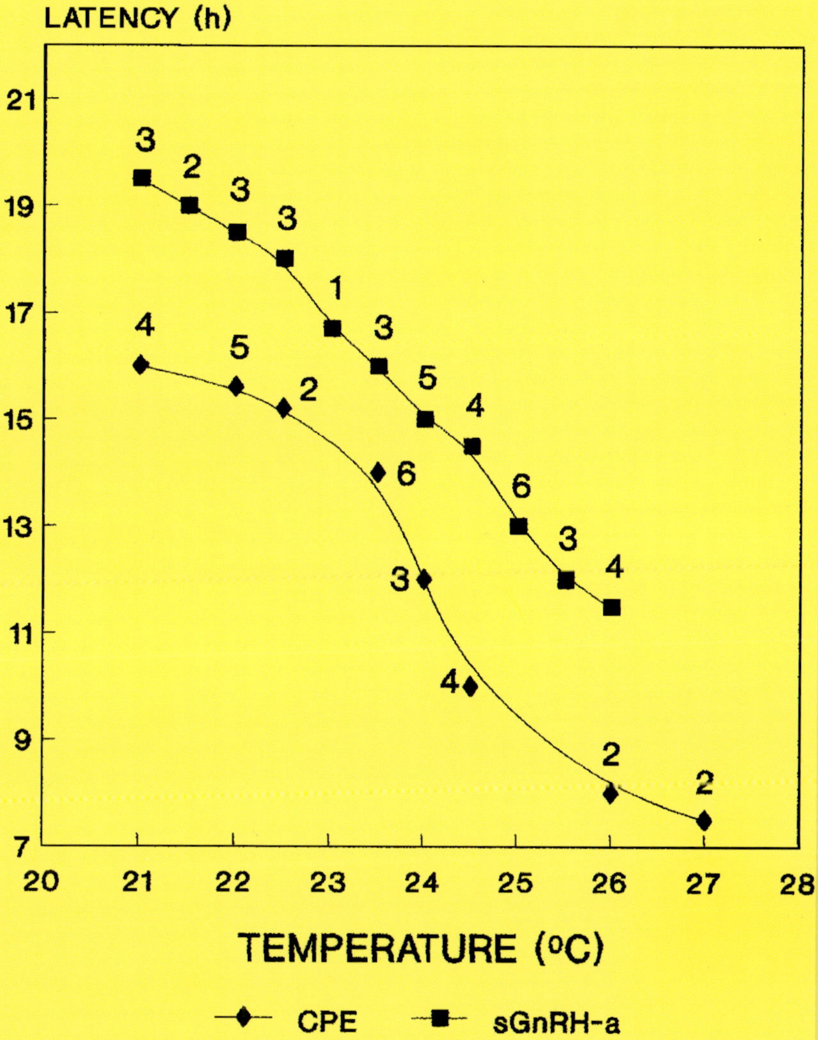

Figure 12. The places on a female/male body suitable for injection of hormones to induce spawning in koi: (a) intramuscular, (b) behind the dorsal fin, and (c) peritoneal injection.

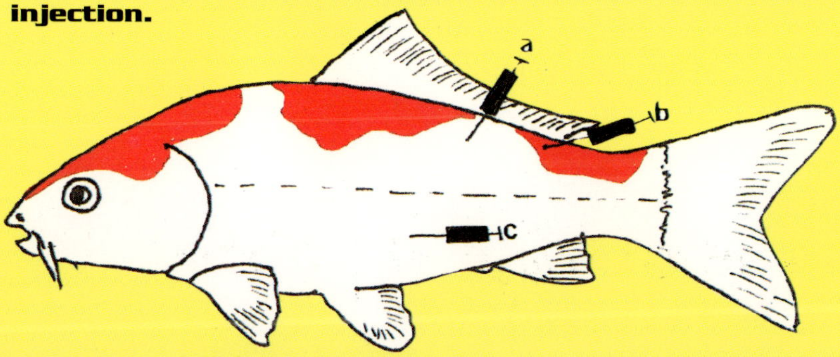

FERTILIZATION, INCUBATION, & EMBRYONIC DEVELOPMENT

COLLECTION OF KOI GAMETES AND FERTILIZATION

In many fish species, including the koi, the so-called **dry method** of fertilization is commonly used. The ripe and ready-to-spawn females are gently transferred in a soft net from the holding tank to a smaller container filled with tranquilizing solution (e.g. 100 ppm Benzocaine®) for anesthesia. After a few minutes the calmed female is carefully wiped and dried with a soft towel. To avoid an unwanted release of eggs by the female at handling, one of the co-workers clogs the genital opening with a finger. The female must be completely dry at stripping to avoid pre-activation of the eggs before the addition of sperm and fertilization. The eggs are stripped by gentle squeezing of the swollen belly, which results in a flow of ovulated eggs into a clean, dry bowl .

The number of stripped eggs can be estimated by weight. Although young and small sized (<1kg) females tend to produce relatively small sized eggs, a koi female at the preferable spawning size of 2-3 kg may produce about 500 grams of eggs (one gram contains about 600-700 unswollen eggs). A gonado-somatic index (GSI) of 20-30% is quite common for a ripe koi female. It should be noted, however, that some koi varieties, as the **ohgon** or **shusui**, are likely to be more prolific than other koi variants. The relative fertility (**RF = the quantity of eggs per kg female**) of females is

Figure 13. Wiping of a wet and tranquilized koi spawner before stripping.

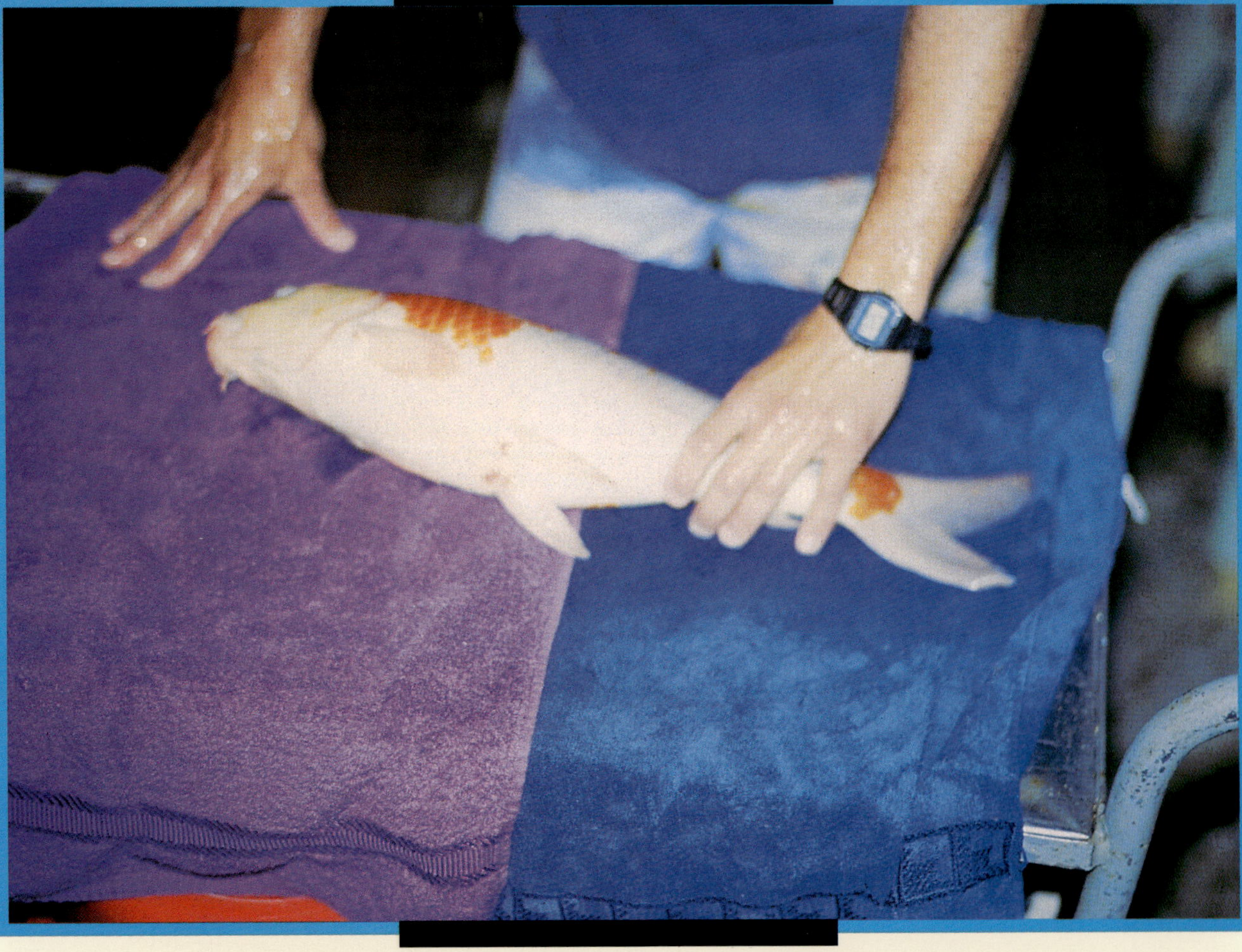

Figure 14. Stripping of eggs into a bowl.

age and size dependent, e.g. the RF of young and small females will be very high since they produce a large number of small size eggs.

The egg (ovum, germ cell) is protected by a soft shell (chorion) that encloses the perivitelline space, which is filled with vitelline fluid. The yolk is retained inside the space within a membrane. An opening in the chorion, the micropyle, provides the only entryway for the penetrating spermatozoon. The pre-activated koi eggs can be efficiently stored at moderate and stable temperatures (19-21°C) for short durations (5-8 hours) before fertilization (Rothbard *et al.*, 1996). As soon as the egg is fertilized and its shell hardens, the micropyle closes and no other sperm can penetrate and fertilize the egg.

Koi males are treated in the same way as the females. After wiping and drying the fish to avoid premature activation of sperm, milt can be stripped and collected in 100 ml glass beakers. Koi sperm can be diluted in 0.9% NaCl (1:3; sperm:saline solution) and stored for 7-10 hours, or even for 24-30 hours in a home refrigerator (at 4-6 °C) until used. However, it is recommended that the motility of the sperm is examined prior to use. Examination of sperm motility is a simple procedure; a drop of sperm is mixed with a drop of tap water on a microscope slide and checked under the microscope at a magnification of 100-400X. Non-activated sperm is motionless and the addition of water immediately activates the spermatozoa, which become motile and start moving very fast. Spermatozoa movement can last only 1-2 minutes due to their limited energy content.

To avoid inbreeding and to ensure a good fertilization rate, sperm from 2-3 males can usually be pooled before use. About 5-8 ml of saline diluted sperm is required to fertilize one kg (600,000-800,000) of eggs.

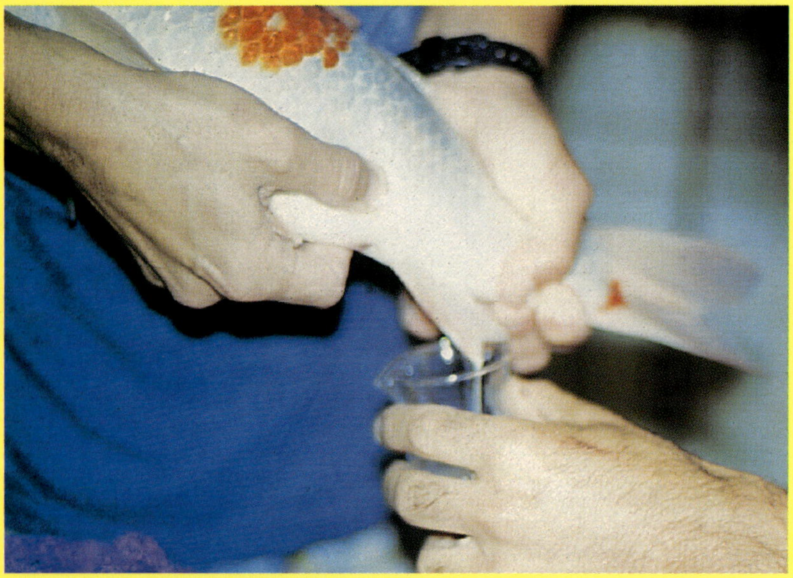

Figure 15. Stripping a koi male. The sperm is collected in a small glass beaker.

HOW TO REMOVE THE STICKINESS OF KOI EGGS

As in many fish species, and particularly in cyprinids, koi eggs are covered by a thin layer of glycoproteins that become adhesive with the activation of water. The role of stickiness in nature is to make possible the attachment of activated eggs to immersed aquatic vegetation. In hatcheries, this sticky protection is removed from the fertilized eggs prior to their incubation in running water. The so-called **degumming** of eggs is accomplished through continuous rinsing with solutions comprised of different proportions of salt and urea, a method developed by Woynarovich, a famous Hungarian aquaculturist. Because the first solution slows and prolongs sperm activity it is also sometimes called the **fertilization solution** (Woynarovich & Woynarovich, 1980). An addition of 10% milk to the solution can improve the entire process of the removal of stickiness. The entire process of degumming, during which the eggs must be mixed every 5-10 minutes and the solution replaced by a fresh one, lasts about an hour (**Table 5**). In some large hatcheries a shaking table facilitates the rinsing by treating 20-30 batches of eggs at once (see p. 36). At the end of the rinsing, the degummed and hydrated (swollen) eggs are quickly

Figure 16, 17. Dry fertilization. The sperm is poured onto the eggs.

washed with tannic acid (tannin) solution, which eventually denatures the glycoprotein layer. The non-sticky eggs can be placed in incubators without the danger of clumping.

EMBRYONIC DEVELOPMENT

When compared to the eggs of cold water fishes (e.g. salmonids), or even to other cyprinids (e.g. grass carp), the embryonic development of koi eggs is rather fast. The duration of incubation is temperature dependent (see graph p. 38) and at 23-24°C the yolk-sac larvae will hatch after 50-60 hours.

As soon as hydration of the eggs has been accomplished, the zygotes start their embryonic development. The egg nucleus starts its first mitotic cleavage, ending with

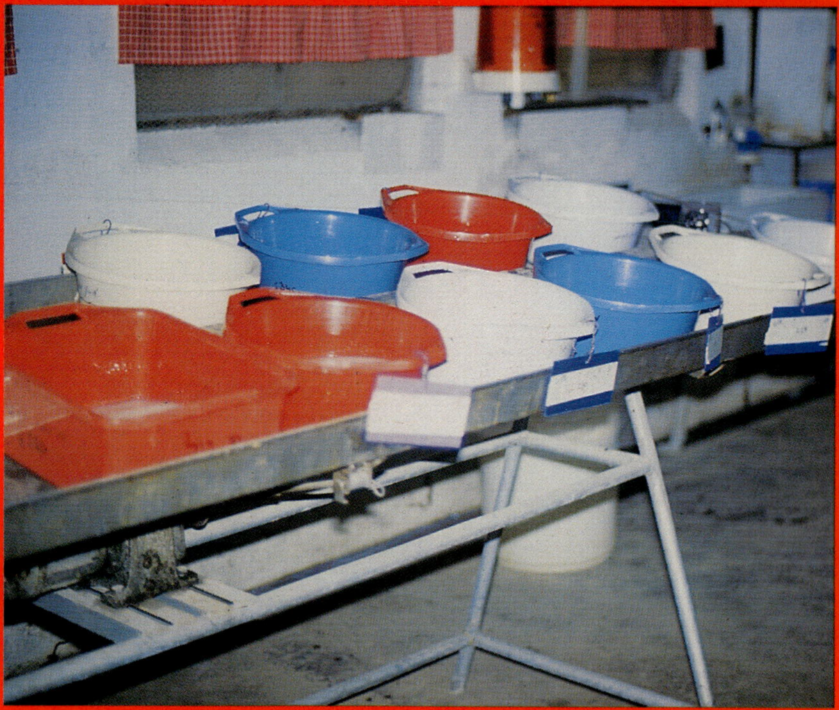

Figure 18. A table for simultaneous rinsing and degumming of many batches of eggs.

TABLE 5

Treatment of koi eggs (washing and rinsing) to remove the stickiness before artificial incubation.

SOLUTION	DURATION OF RINSING	MODE OF RINSING	PURPOSE
Fertilization (Woynarovich) solution			
3 g NaCL 4 g Urea 1 L water	4-5 min.	Continuous mixing.	To prolong the fertilization. Initiation of egg hydration and degumming.
Modified Woynarovich solution			
3 g NaCl 20 g Urea 1 L water (=200 ml milk)	60-80 min.	Mix every 5-10 min. and replace the solution.	Continuation of eggs swelling (hydration) and degumming.
Tannic acid solution			
1 g tannin 1 L water	Few seconds and washing with water.	Dipping the eggs.	Hardening the egg shell.

At the end of rinsing, the number of eggs is volumetrically estimated before incubation (1 ml eggs=80-100 eggs)

Below: Shiro Utsuri.

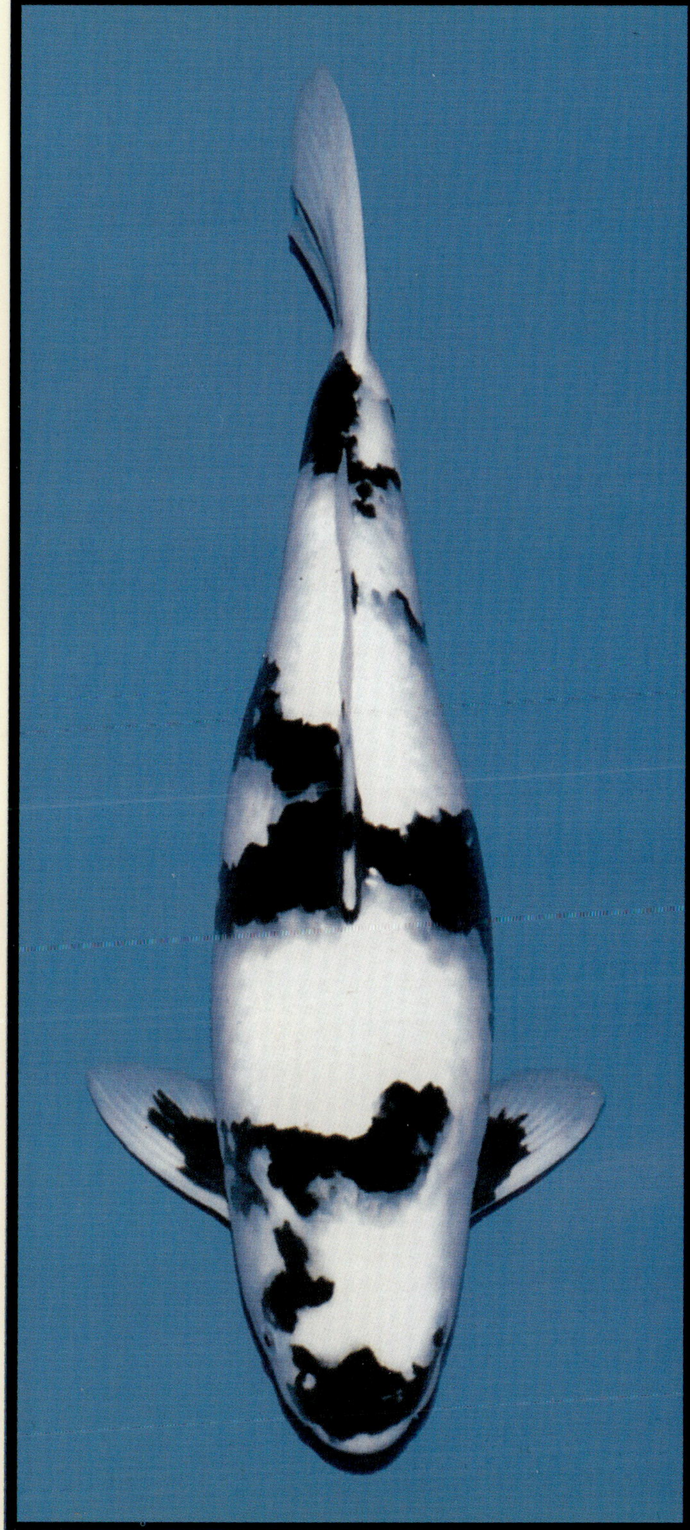

Above: Showa Sanshoku.

two identical cells, which are products of the first mitotic division. These further divide (double) into 4, 8, 16, 32, and more cells. The process can be easily followed under simple binocular (40X) magnification. After a few hours of development, the group of cells looks like a mulberry (**morula**) that settled on the yellow yolk mass. The koi (and other fish) embryos are very sensitive to handling during the morula stage and must be provided with continuous movement and sufficient aeration. The next developmental phase is distinguished by a single cell (**blastomere**) layer, termed the **blastoderm**, which covers the yolk, forming the so-called **blastula**. An opening on the blastula will later develop into the primary mouth (**blastopore**). At this stage the embryo has already developed some cell layers and is now called the **gastrula**. This particular stage of development is also very critical, because the embryo ends its synchronous cell division and the cells start to differentiate into various tissues. Subsequently, **optical vesicles** can be identified at the front of the embryo's head. Together with the development of blood vessels, the heartbeats can be easily seen under a binocular microscope. A few hours later the embryo starts to twitch its tail, which is followed by convulsions of the entire body. The embryo moves more and more vigorously inside the perivitelline space. At this stage certain enzymes produced by the embryo start to act on the egg shell from the inside, helping to break it and allowing the larva to hatch out. The larvae have a yolk-sac extended along their abdomen. A schematic presentation of the

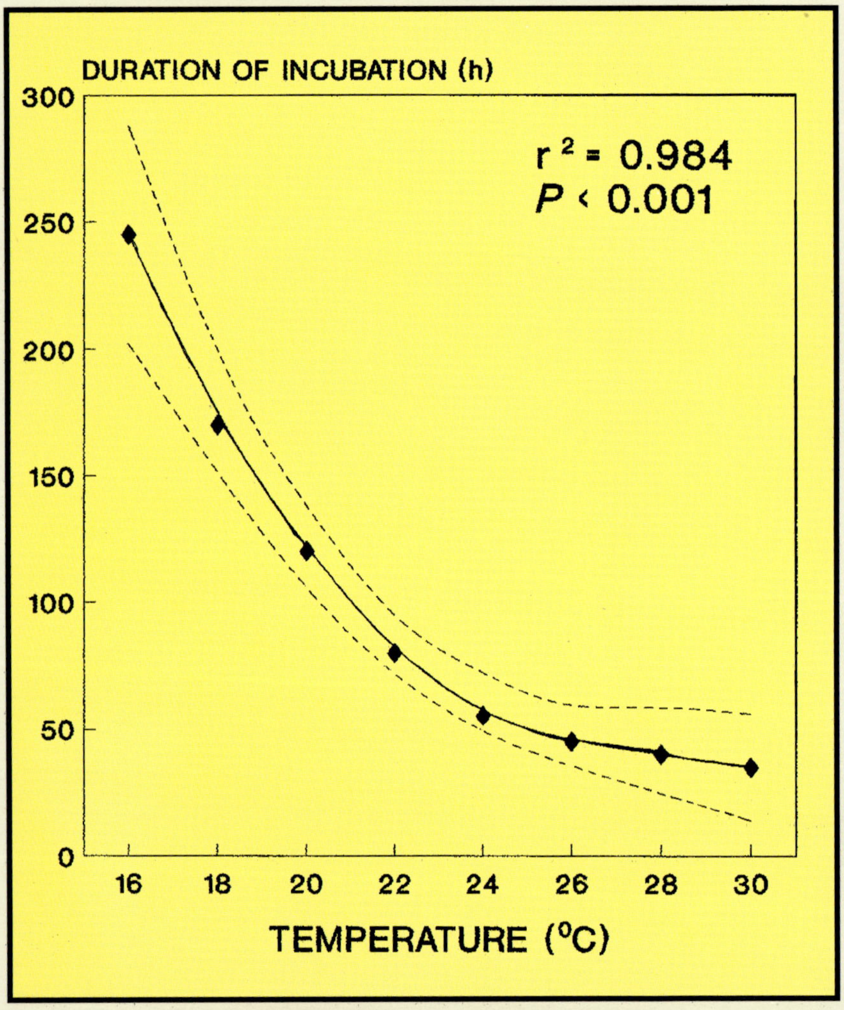

Figure 19. The relationship between the temperature and the duration of incubation of koi eggs.

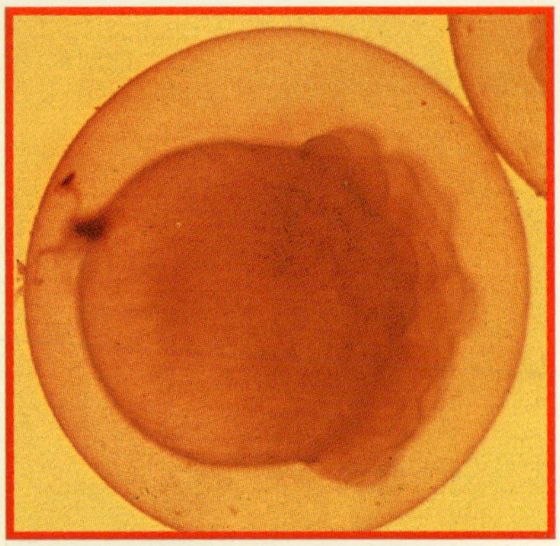

Figure 20. The 8-celled embryo.

Figure 21. Koi embryo before hatching.

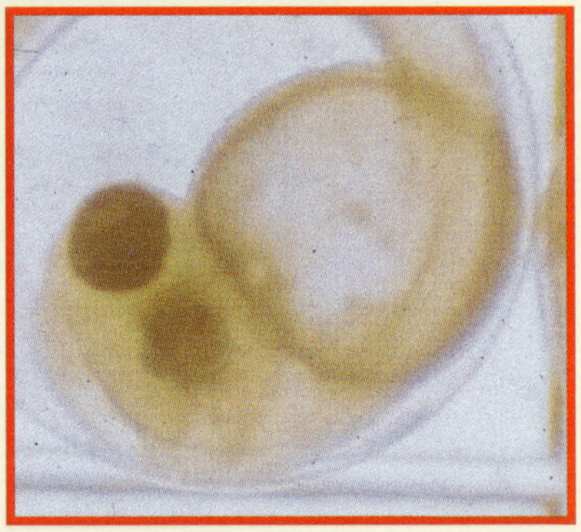

Below: Taisho Sanke.

Above: Kohaku.

progressive stages in the embryonic development of koi are shown on p. 41.

The incubation of koi eggs is carried out in conical incubators of various shapes see p. 42. The eggs are moved by water flow, which enters the incubator via an inlet attached to the sloped bottom and drained by an upper outlet. After hatching, the larvae swim with the current into the adjacent larvae tanks. The passage of the free-swimming larvae from the incubators to the larvae tanks can last for 2-3 days.

THE LARVAL STAGE

Hatching starts with the

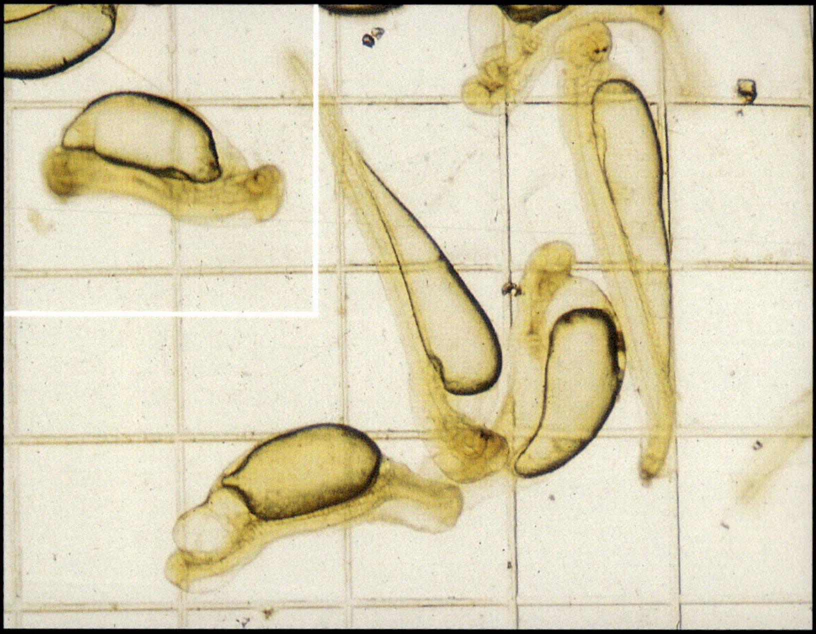

Above: Figure 22, 23. Hatched koi embryos. Below: Figure 24. A group of koi embryos before hatching.

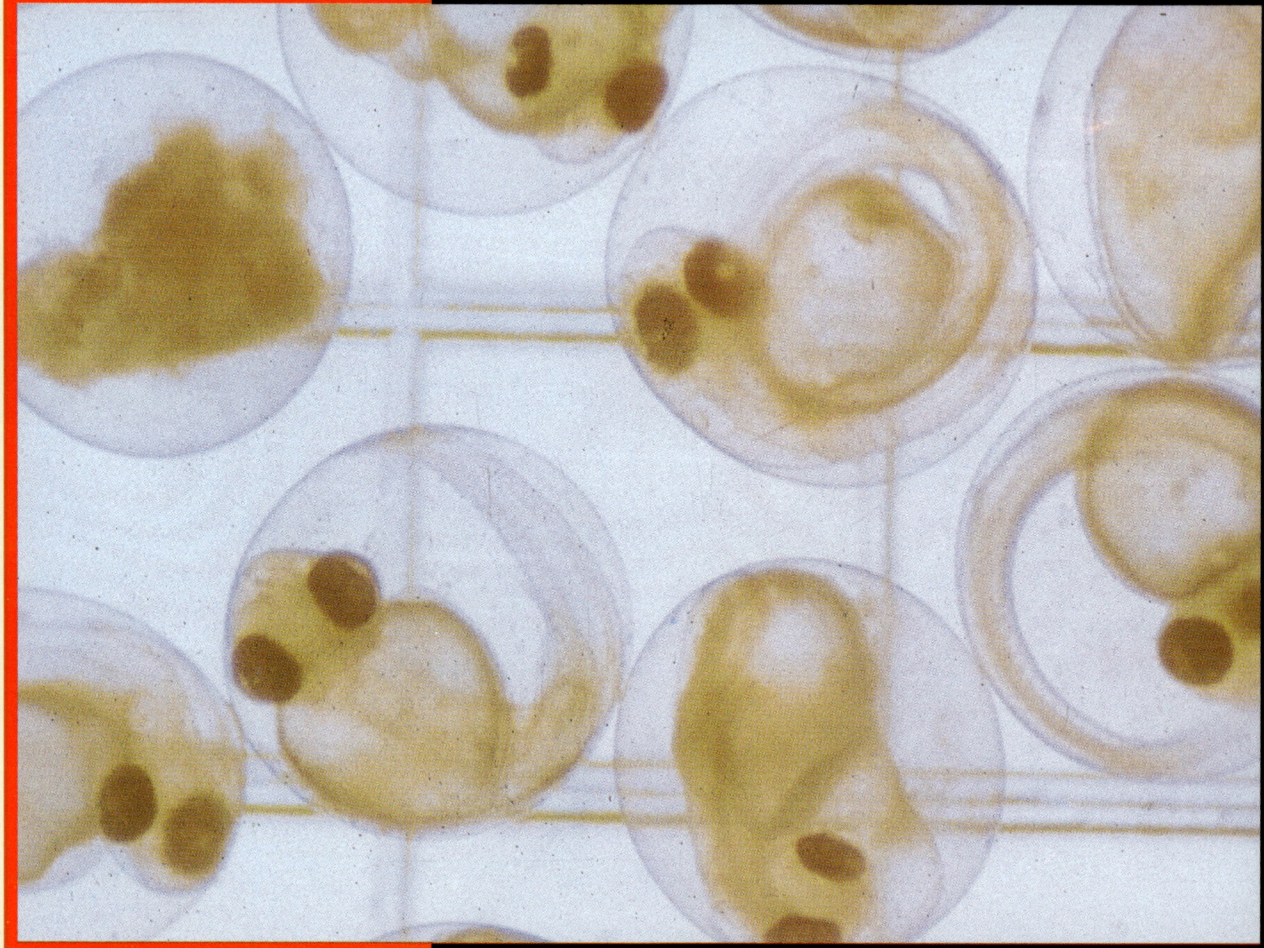

A scheme of embryonic development in koi:

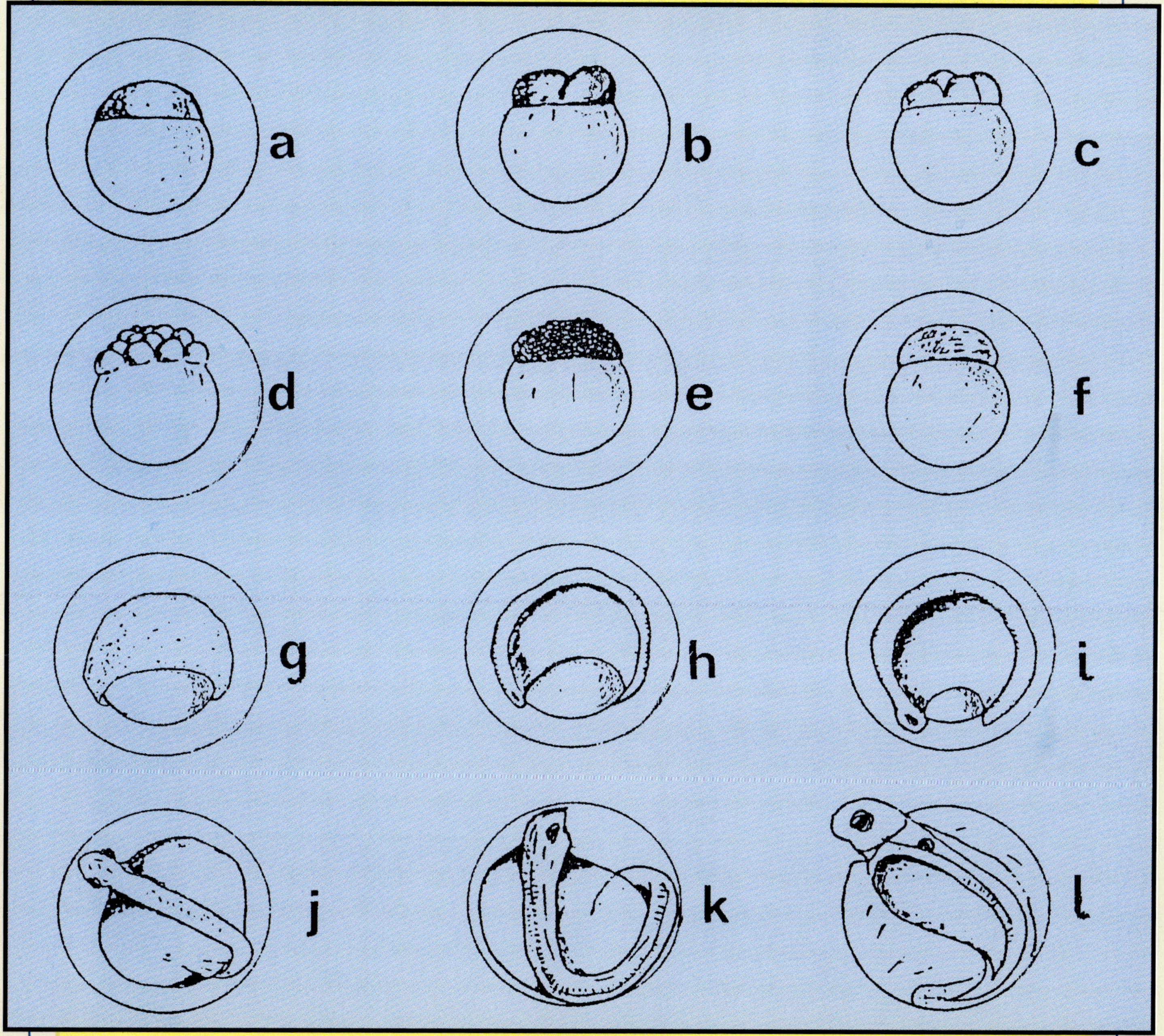

Figure 25. a. The fertilized ovum.
b. The two-celled stage.
c. The four-celled stage.
d. The early morula.
e. Morula comprised of many cells.
f. Blastula.
g. The beginning of gastrula formation.
h. The blastopore.
i. Formation of the gastrula.
j. Initiation of embryonic development.
k. The moving embryo.
l. The hatching larva.

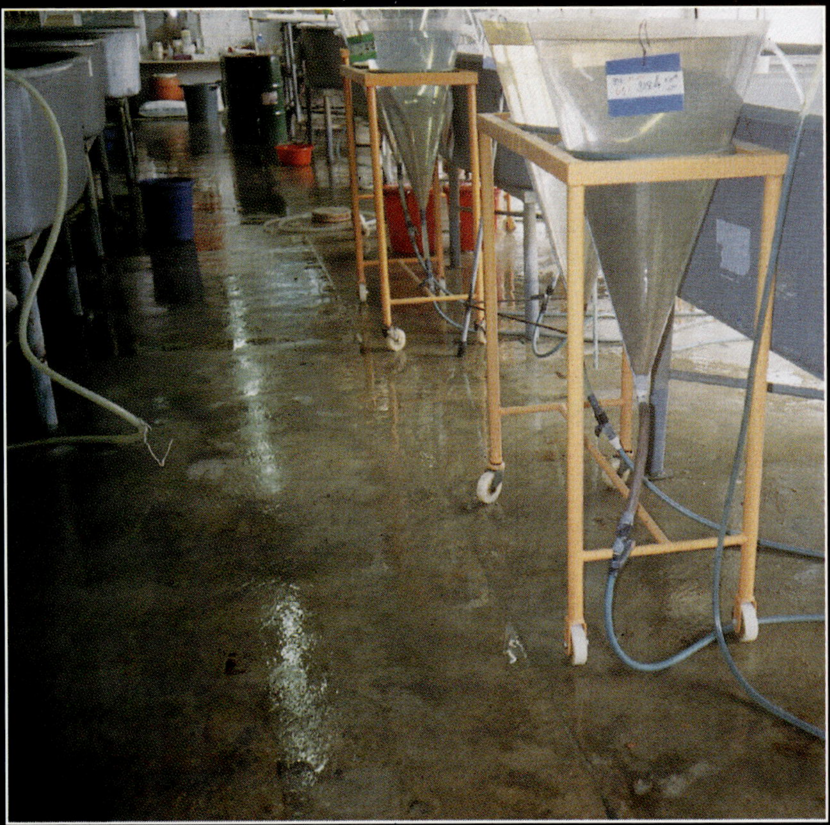

Figures 26, 27, & 28. Conical (Zuger) incubators for koi and carp eggs.

gradual weakening of the egg shell due to the effect of endogenous enzymes produced by the pre-hatching embryo. The optimal hatching temperature for koi larvae is between 22° and 25°C. At lower temperatures the reduced metabolic rates, including the enzymatic activity of the embryo, will prolong the hatching. In contrast, at the highest temperatures the larvae may hatch out before the embryo has completed its development. Both extreme conditions may result in the mortality of the larvae.

The hatched larvae still do not possess a swim bladder and therefore are unable to regulate their buoyancy. This is the reason that the larvae swim vertically (fry swim horizontally), twisting their tail. The larvae attach themselves, for frequent resting, to the tank walls or to any immersed substrates. At this sensitive stage the aeration in the tank must be sufficient but not vigorous in order to avoid loss of energy by the fragile and sensitive larvae.

The yolk serves as the sole source of energy for the larvae for the first 3-4 days after hatching. At hatching, the larvae have no mouth opening. Their stomach, gills, and swim bladder are not developed yet. These will develop during the first days after hatching, utilizing the food accumulated in the yolk. With the absence of gills, the gaseous exchange is executed by diffusion through the entire body surface, and therefore koi larvae are extremely sensitive to oxygen depletion and osmotic pressure of the aqueous environment. Oxygen depletion can be localized in "dead corners" of the larvae tank. Therefore, circular tanks with efficient circular water currents and aeration are of a preferable form.

The gill filaments develop two days after hatching and

Figure 29. Circular 1.5 m³ larvae tanks with attached incubators.

at this time the larva is able to swallow a small bubble of atmospheric air to fill up the gas bladder. Simultaneously with the filling of the gas bladder, the larva develops its sense of hydrostatic balance and starts to swim horizontally. At this early stage of development the larva still does not posses paired fins, but only a single, elongate fin around its entire body. Three-day-old larvae actively start seeking food. This is the time that the larva has transformed into a fry.

THE PRIMARY NURSING OF KOI LARVAE

As soon as the yolk sac has been absorbed, the larvae start seeking an external source of food. At this stage the availability of food and its quality are crucial factors for normal development of the small koi fry. Since the sensory organs (e.g. the senses of smell and taste) are not developed as yet in the larvae, their capability to collect food exclusively depends on their visual sense and sufficient illumination is important. Since the koi larvae have a well structured cone retina they need high intensity illumination for feeding. Insufficient light will significantly reduce the ability of the larvae to feed.

In a pond, koi fry prey on tiny aquatic organisms that have to be present in adequate densities and relatively close to the larva's mouth. Planktonic organisms, at sizes of 100-500 μm (fresh water rotifers and copepods), comprise the most suitable food for the first days of fry feeding.

An excellent live food for koi fry are nauplii of the brine shrimp, *Artemia salina*. These are available from almost all dealers and suppliers of aquarium accessories, and advertised in many aquarium hobbyist magazines (including *Tropical Fish Hobbyist* magazine). However, those koi hatcheries that are involved in mass production of fry are advised to contact large and certified producers of *Artemia* cysts.

The simplicity of obtaining live *Artemia* nauplii from dry, stored cysts at any time and in any quantity, made them an approved food for the first feeding stages of many fish species. Due to their importance for aquaculture and their biological features, brine shrimp have been extensively investigated. The world famous institution involved in nutritional research and the introduction of new genetic strains of *Artemia*, is the **Artemia Reference Center (ARC)** of the University of Ghent (Belgium). The reader is also advised to refer to works on decapsulation of *Artemia* cysts, which increases and improves the hatching rates of *Artemia* nauplii (Sorgeloos, *et al.*, 1977).

Above and Right: Figures 30 & 31. A battery of incubators for Artemia eggs.

At the age of 10-14 days, koi fry that were maintained in the hatchery are strong enough and can be stocked in earthen ponds for secondary nursing. At densities of 800,000-1,000,000 fry/ha, koi fry can reach a size of 4-6 cm in 5-6 weeks. When stocked at these densities, the first culling of koi fry can be attempted at the age of 3-4 weeks and at the initial size of 2 cm. At this stage successful production of koi depends largely on proper preparation of the nursing pond, as well as on nursing methods and management.

HATCHERY SANITATION & LARVAL HEALTH

It is beyond the scope of this book to deal in details of fish health. There are numerous periodicals and books on fish hatchery management and fish diseases (many of these have been published by T.F.H. Publications, Inc.). Among these the excellent *Textbook of Fish Health* by Dr. George Post (1987) is highly recommended. This part of the book deals with some problems of sanitation in the koi hatchery, as well as parasites and diseases that may cause heavy losses, or even the total mortality of koi larvae and small fry.

SANITATION IN THE HATCHERY

Usually the outbreaks of diseases in koi hatcheries, as well as in other aquaculture systems, are caused by the presence of at least two factors, (a) presence of a parasitic organism and (b) an abnormal situation in the affected fish (results of stress and injuries caused by poor handling, starvation or poor nutrition, bad water quality, etc.). In fish hatcheries the strict and routine control of hygiene and sanitation is a prerequisite. Usually, treatments are performed with drugs and chemical additives dissolved in the water. These treatments can last between a few seconds or a few minutes (**dips**), immersions of fish applied for hours (**baths**), and chemicals that are added to running water and applied for hours (**flushes**). The tremendous list of drugs and medicaments used for sanitation and disease control can be found in every textbook on fish disease prevention. The treatments are commonly carried out at three levels:

Prevention—It should be always kept in mind that the fish introduced to the hatchery from outside (e.g. broodstock) are passive carriers of parasites. Therefore special attention must be paid to fish introduced from outside the facility, as for instance in the case of purchasing new broodfish from other sources. Larvae tanks, nets, buckets, and containers, as well as all other equipment used in the hatchery, must be disinfected from time to time. The biofiltration, in case of a closed and recirculating water system, even equipped with an ozonizer or UV-sterilizer, must be frequently checked to prevent the establishment of an undesired population of bacteria.

Prophylaxis—Since environmental factors, as pH, temperature, and water hardness, may affect chemical treatments and increase the toxicity of chemical compounds, the chemotherapy of fishes is a very complicated procedure and no standards have been yet established. However, the most important rule should be to empirically examine the treatment on a representative sample before application of chemotherapy to large stocks of fish. The results and data collected from such trials should be recorded and analyzed for further use and before massive application.

Treatment—Fish treated with any drug by any method (dipping, bathing, flushing, or administration by injections), should be kept starved for 24-48 hours prior to treatment. Starvation is also indispensable before packing fry for a prolonged transportation. The aim of starvation is to reduce the metabolism of the fish to a minimum (e.g. oxygen consumption, ammonia production, etc.).

For a complete list of diseases and their appropriate treatments, the reader is referred to an extensive collection of books and articles available on the shelf in every book store and library. The reason for describing the gas bubble disease in detail in this publication is due to its common outbreak in fish hatcheries and its fatal consequences to koi larvae.

THE GAS BUBBLE DISEASE

The gas bubble disease (GBD), or gas embolism, belongs to the category of diseases caused by the physical factors that affect the aquatic environment of fishes. Fishes, as other living organisms, may subsist within a quite narrow range of environmental limits (e.g. temperature, salinity, pH, oxygen level, hydrostatic pressure, etc.). Actually, gas bubble disease is the result of a physical process that results from an unbalanced hydrostatic pressure of dissolved gases in the water

Figure 32. Small koi larvae affected by gas bubble disease.

or a sudden increase in water temperature. For instance, gases are more soluble in cold water, but the reverse is true with an increase of temperature, since then their solubility decreases and the gases tend to supersaturate the water.

The koi (and other fish) larvae are exposed to GBD danger until the soluble gases in the water achieve equilibrium with the atmosphere. A sudden and rapid increase in water temperature or changes of hydrostatic pressure upset the hydrostatic balance and result in a supersaturation of dissolved gases. These diffuse into the fragile larval tissues and provoke GBD. For instance, high and unstable hydrostatic pressure, which usually exists in water pipelines, will also result in supersaturation of the atmospheric gases in the hatchery water flow and, finally, with GBD infestation. Consequently, the appearance of GBD can be predicted (and avoided) by routine examination and analysis of the total gases dissolved in the water.

Gas bubble disease occurs when the dissolved gases (O_2, N_2, CO_2, and others) come out of the water in various fish tissues and appear as small gas bubbles that can be seen with the naked eye, i.e. without optical magnification. These usually tend to accumulate in loose connective tissues (close to the eyeballs) or along the lateral line. Small gas bubbles are often present within the gill capillaries (embolism). The presence of gas bubbles can block blood vessels (hemostasis) or tear tissues and, finally, result in the mortality of the fish. Chronic GBD will appear in water that is saturated by more than 105-110% of atmospheric gases. In holding tanks crowded with koi larvae and exposed for hours to an excessive saturation of gases, GBD will result in total loss of the larvae.

REFERENCES

Drori, S., Ofir, M., Sivan, B. & Yaron, Z. (1994) Spawning induction in common carp (*Cyprinus carpio*) using pituitary extract or GnRH superactive analog combined with metochlopramide: analysis of hormone profile, progress of oocyte maturation and dependence on temperature. *Aquaculture*, 119:393-407.

Hepher, B. & Pruginin, Y. (1981). *Commercial Fish Farming (with Special Reference to Fish Culture in Israel)*. John Wiley & Sons, New York.

Horvath, L. (1985). Egg development in the common carp (*Cyprinus carpio* L.). pp. 31-77 In: *Recent Advances in Aquaculture*, Vol. 2. (J.F. Muir & R.J. Roberts, eds.). Westview Press, Colorado.

Horvath, L. (1986). Carp oogenesis and the environment. pp. 109-117 In: *Aquaculture of Cyprinids.* (R. Billard & J. Marcel, eds.). INRA, Paris.

Ihering, R.V. (1937). A method for inducing fish to spawn. *Prog. Fish Cult.*, 34:15-16.

Kabata, Z. (1985). *Parasites and Diseases of Fish Cultured in the Tropics.* Taylor & Francis, Ltd., London.

Levavi-Zermonsky, B. & Yaron, Z. (1986). Changes in gonadotropin and ovarian steroids associated with oocyte maturation during spawning induction in the carp. *Gen. Comp. Endocrinol.*, 62:89-98.

McFarland, W.N., 1960. The use of anesthetics for the handling and the transport of fishes. *California Fish and Game*, 46(4):407-431.

Rothbard, S., Rubinshtein, I. & Gelman, E. (1996). Storage of common carp, *Cyprinus carpio* L., eggs for short durations. *Aquaculture Research*, 27:175-181.

Rothbard, S. & Wohlfarth, G.W. (1995). Methods for improvement of Japanese ornamental (koi) carp. *Tropical Fish Hobbyist*, 43(No. 7, # 469):224-226, 228-232, 234-242.

Sorgeloos, P., Bossuyt, E., Lavina, E., Baeza-Mesa, M. and Personne, G., 1977. Decapsulation of *Artemia* cysts: A simple technique for the improvement of the use of brine shrimp in aquaculture. *Aquaculture*, 12:311-315.

Wohlfarth, G. (1977). Shoot carp. *Bamidgeh*, 29:35-56.

Woynarovich, E. & Woynarovich, A. (1980). Modified technology for elimination of stickiness of common carp (*Cyprinus carpio*) eggs. *Aquacultura Hungarica (Szarvas)*, 2:19-21.

Yaron, Z. & Levavi-Zermonsky, B. (1986). Fluctuations in gonadotropin and ovarian steroids during the annual cycle and spawning of the common carp. *Fish Physiology and Biochemistry*, 2:75-86.

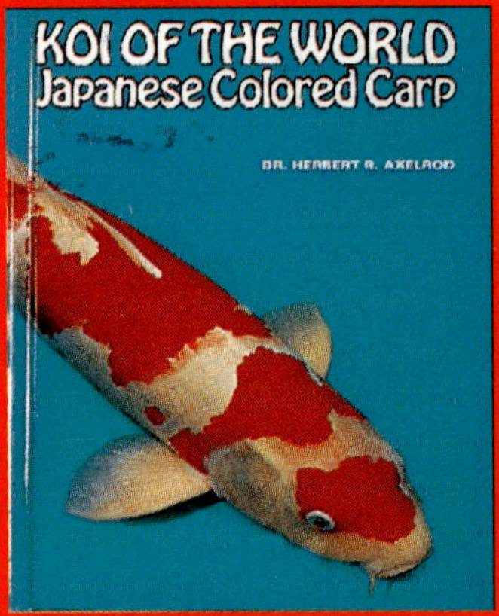